Praise for Whole System Design:

Speaking recently, I outlined what I thought were the requirements for the engineer of tomorrow. I was quickly corrected. Today's engineer needs to be engineering with tomorrow already clearly in mind. This book encourages and leads today's engineer on a journey to meet tomorrow's needs. Systems thinking and asking the right questions opens up far more design options and solutions than we first expect. And some of those solutions bring the breakthrough improvements that go far beyond the incremental. Like many books, this one seems a little too simple at first, but I challenge the reader who feels that way to jump to the back and look at the examples. Then go back and read again. There is real power in its simple approach. Engineers are often caught up in looking for the incremental improvement, but I would suggest that our current challenges need more than that. I'd encourage all engineers to look at this book. Dip into it at first, then, come back to it. There is an elegance in the approach it advocates. I had a design lecturer once who commented that I had correctly answered the question, but that I might have done better by asking a very different question. I think he would like this book.

Martin Dwyer, Director, Engineering Practice and Continuing Professional Development (CPD), Engineers Australia

Whole System Design is a comprehensive resource to support professional, academic and student engineers in complex problem-solving around sustainability – an area of focus recommended by the 2008 Review of Engineering Education in Australia, *Engineers for the Future*. As the book shows, engineers and designers can make a significant difference to the current global environmental crisis by reducing environmental impacts in the design phase of a wide range of projects.

Associate Professor Roger Hadgraft, Director, Engineering Learning Unit, Melbourne School of Engineering, and President of Australasian Association for Engineering Education, Australia

The Natural Edge Project's *Whole System Design* book will provide a valuable resource that can contribute significantly to the technical design curriculum in university courses and professional training. I have used a Whole System Design approach, as is described and demonstrated in this book, to improve resource efficiency of products and industrial processes often by a factor of 2 or better. An exciting consequence of applying a whole system design approach is the drastically reduced need for end-of-pipe treatment, both in the local area and potentially in the wider air, soil and waterways. This book is the first resource that I've seen that goes into sufficient detail for the reader to comprehensively grasp the concepts involved in a Whole System Design approach. A great attribute of the book is that it is not simply a set of a stand-alone ideas – it provides a strong foundation for embedding sustainable design into the popular design process already taught to students and professionals in Australia and around the world. It is evident that a great deal of thought went into ensuring that the ideas in the book could be quickly and easily integrated with current practices, and ensuring that the ideas are universally applicable to all engineering and technical design disciplines. I commend The Natural Edge Project for their efforts and the Department of the Environment and Water, Heritage and the Arts for supporting the project.

Adjunct Professor Alan Pears, School of Global Studies, Social Science and Planning, Royal Melbourne Institute of Technology, Australia

Whole System Design underpins efforts to help get our societies onto sustainable pathways. This book is a much-needed contribution providing, in detail, instructions on how to implement sustainable design for green buildings, more eco-efficient products, ICT systems and fuel-efficient cars, to help us build healthy cities.

Dr Steve Morton, CSIRO Group Executive, Manufacturing, Materials and Minerals, CSIRO, Australia

Climate change poses a significant challenge but also a great opportunity. Mitigating climate change successfully will involve transforming our energy systems. As part of this transformation, it is vital that existing technologies and designs are re-examined to identify new ways to make them more energy efficient. The Whole System Design approach presented in this book offers engineers an advanced strategy to enable them to achieve large energy efficiency savings. We urge you to read and absorb the book's whole system design framework and then see how whole system design can be applied to achieve large energy-efficiency savings in the book's detailed technical case studies.

Dr John Wright, Director, CSIRO Energy Transformed Flagship, CSIRO, Australia

Whole System Design (WSD) developed by The Natural Edge Project (TNEP) will be an invaluable resource in the near future for the education of systems engineers on matters of sustainability and design. It provides a seamless link between the traditional system engineering design approach and the wider perspective of environmental and social effects that future engineers need to consider. The WSD material is lucid and concise but also has sufficient technical depth to be useful and challenging for all students in the tertiary sector. In particular, the high impact examples and case studies clearly illustrate the new systems thinking. I am already integrating the WSD book into the systems engineering curriculum of the ANU Engineering undergraduate programme and the impact, in terms of sustainability awareness and responsibilities for future engineer practice, is immediate. The TNEP material is, therefore, already changing the perspective and thinking of our future engineers and aligning their design skills to address the global environmental challenges.

Dr Paul Compston, Department of Engineering, Australian National University, Australia

We all have a major role to play in reinventing our business model and shaping our future, whether we are engineers, designers, governments, business people or entrepreneurs ... small, simple steps won't cut it to deal with the major global challenges of climate change and environmental degradation we are all facing. There are thousands of cases that demonstrate that, yes, we can transform these challenges into the foundations of a more sustainable, profitable, and desirable societal model. But where to start? What is the most effective, profitable and desirable way to implement the change we want to see? *Whole System Design* provides essential, hands-on guidance to kick-start this next industrial revolution. This book moves the reader from thinking 'hmmm ... this is interesting' to 'I'm gonna do this!' It reframes the future not as fate, but as choice. A choice each one of us can define, prioritize and execute.

Professor Serge de Gheldere, Founder and Managing Director of Futureproofed and Guest Professor and Director at Group T University College Leuven, Belgium

The book *Whole System Design* is a clever feat of engineering that bridges the traditional divide between technological and design thinking. It shows how we can cross the giant chasm between conventional and sustainable systems in small, easy steps – provided we start now. It should be read by all engineers as a matter of urgency.

Professor Janis Birkeland, School of Design, Queensland University of Technology, Australia

Whole System Design gives a comprehensive introduction to whole system design approach as the basis for transformative action. Education for Sustainability has to be more than 'bolt on' environmental papers in existing programmes, and this is the best example I've seen of resources to support sustainability as an integrated and transformative driver.

Associate Professor Samuel Mann, Department of Information Technology, Otago Polytechnic, New Zealand

The Industrial Pumping Systems Chapter is a nice example that illustrates the point well.

Emeritus Professor Bruce R. Munson, Department of Aerospace Engineering, Iowa State University, USA

The Chapter on Domestic Water Systems within Whole System Design developed by TNEP eloquently captures the current household water challenge; that is, achieving both fit-for-purpose and efficient water use, to reduce the water footprint of this sector of the economy. Current data about water consumption, available technology, and cost across the life cycle of the technology illustrate sensible, simple and appropriate design solutions for engineers looking to understand and implement best-practice water systems engineering. Capital and operating costs are included by TNEP through case studies, to confirm that water-efficient design is the only way forward to meet water needs for households, on a least-cost basis, and a quality appropriate to purpose. In addition, the chapter will enlighten users on the environmental and economic benefits of moving from linear household water use, treatment and disposal systems, to more enclosed water-use systems, through appropriate and sensible engineering design.

Nick Edgerton, AMP Capital Sustainability Fund, formerly of the Institute for Sustainable Futures at the University of Technology Sydney, Australia

Whole System Design
An Integrated Approach to Sustainable Engineering

Peter Stasinopoulos, Michael H. Smith,
Karlson 'Charlie' Hargroves and Cheryl Desha

London • Sterling, VA

First published by Earthscan in the UK and USA in 2009

The views and opinions expressed in this publication do not necessarily reflect those of the collaborating parties: Australian Government; Australian Federal Minister for the Environment, Heritage and the Arts; Australian Federal Minister for Climate Change and Water; United Nations Educational, Scientific and Cultural Organization; and the World Federation of Engineering Organizations. While reasonable efforts have been made to ensure that the contents of this publication are factually correct, these parties do not accept responsibility for the accuracy or completeness of the contents, and shall not be liable for any loss or damage that may be occasioned directly or indirectly through the use of, or reliance on, the contents of this publication.

ISBN: 978-1-84407-642-0 hardback
978-1-84407-643-7 paperback

Typeset by Domex e-Data, India
Printed and bound in the UK by MPG Books, Bodmin
Cover design by Andrew Corbett

For a full list of publications please contact:

Earthscan
Dunstan House
14a St Cross St
London, EC1N 8XA, UK
Tel: +44 (0)20 7841 1930
Fax: +44 (0)20 7242 1474
Email: earthinfo@earthscan.co.uk
Web: **www.earthscan.co.uk**

22883 Quicksilver Drive, Sterling, VA 20166-2012, USA

Earthscan publishes in association with the International Institute for Environment and Development

A catalogue record for this book is available from the British Library

Library of Congress Cataloging-in-Publication Data

The paper used for this book is FSC-certified. FSC (the Forest Stewardship Council) is an international network to promote responsible management of the world's forests.

This book is dedicated to Amory B. Lovins and Alan Pears.
To Amory, for his significant contribution to expanding the solution space
for sustainable design and for taking the time to mentor our team,
and to Alan for sharing with us his enthusiasm, insights and lessons learnt
from a life dedicated to whole system design.

Contents

List of Figures and Tables

Figures

Tables

Forewords

I

Many of the systems currently in place are not very environmentally sustainable or cost effective in terms of their utilization and the associated costs of operation and support. System performance requirements (and the system's ultimate impact on the operational environment) rarely meet rising customer (i.e. the 'user') expectations for products to be both effective and environmentally benign. The life-cycle costs of most products and technical systems are high. We see symptoms of poor design all around us, manifested in growing problems such as the current environmental crisis. When addressing 'cause-and-effect' relationships, many of these related problems stem from the management and technical decisions made during the early stages of system design and development. In general, the initial requirements for a given system were not very well defined, the system was not addressed in totality (as a 'whole' entity), and a total system life-cycle approach to design for sustainability was not assumed from the beginning. All of this occurred at a critical point early in the system design and development process, and at a time when the results of such decisions would have the greatest impact on the overall effectiveness, efficiency and environmental sustainability of systems in the performance of their intended functions later on.

Given today's environment, there is an ever-increasing need to develop and produce systems that are robust, reliable and of high quality, supportable, cost-effective and environmentally sustainable from a total life-cycle perspective, and that will respond to the needs of the customer/user in a satisfactory manner. Systems in the future must be environmentally friendly, socially compatible and interoperable when interfacing with other systems in a higher-level hierarchical structure. Meeting these challenges in the future will require a more comprehensive sustainable design approach from the start, dealing with *whole systems* and in the context of their respective overall *life cycles*.

From past experience, these objectives can best be met through proper implementation of the *systems engineering process*, or a *whole system approach*, as outlined in this book, to the design and development of sustainable future systems. System requirements must be well defined from the beginning. Systems are addressed in 'total' to include not only the prime mission-related elements utilized in accomplishing one or more mission scenarios, but also the various elements of the system support infrastructure as well. All aspects of the entire system life cycle are considered in the day-to-day decision-making process, including possible impacts on the various phases of system design and development, construction/production, system operation and support, and system retirement and material recycling/disposal. Applicable design characteristics such as reliability, maintainability, human factors, environmental sustainability, supportability, environmental compatibility, quality, economic feasibility (from a life-cycle perspective), etc. must be properly integrated within the design process, along with the required electrical, mechanical, structural, and related parameters.

Proper implementation of systems engineering constitutes a *top-down/bottom-up process*, and not just a bottom-up design-it-now-and-fix-it-later approach. The principles and concepts of whole system approaches to sustainable design outlined in this book are based on the recommendations and experience of leading designers and engineers. Success in applying a whole system approach to sustainable design does require a 'change in thinking' and a slightly different approach in the design and development of future systems.

Implementation of the principles and concepts of whole system design can be applied effectively in the design and development of any type of system, whether addressing communication(s) systems, electrical power distribution systems, mining systems, manufacturing systems, materials handling systems, defense systems, consumer product systems, and the like. In each and all instances, we are dealing with a top-down, whole system and life-cycle approach throughout the initial design and development, and subsequent operation and maintenance phases of the life cycle. The proper implementation of a whole system approach, from the beginning, is essential in meeting the desired goals stated herein.

Based on a review of the content of this book, I sincerely believe that implementing the concepts presented will greatly facilitate accomplishing the objectives defined earlier – that is, leading to the design and development, production and installation of future systems that are robust, reliable and of high quality, supportable and environmentally sustainable, and will be highly responsive in meeting the needs of the customer/user. Of particular interest is the foundation established by the material presented in Chapters 1 to 3. Additionally, the systems engineering process, which is critical in its implementation, is well defined and described in Chapter 3. I feel that following the guidelines presented here will lead to much success in the future.

Finally, I wish to thank The Natural Edge Project for providing me with the opportunity to both review and comment on the material presented within this book, and also for inviting me to be a participant by including this foreword.

Professor Emeritus Benjamin S. Blanchard
Virginia Polytechnic Institute and State University,
Blacksburg, Virginia, USA
October 2008

II

The priorities for the community of engineering professionals, including engineers, technologists and scientists, must necessarily change over the next few years. The rapidly changing world of political, environmental, social and economic challenges demands that we do change and go forward with everything we do.

Engineering professionals must cooperate with other professionals in constructively resolving international and national issues for the benefit of humanity. Engineering professionals around the world understand that they have a tremendous responsibility in implementing sustainable development. Many forecasts indicate there will be an additional 5 billion people in the world by the middle of the 21st century. Supporting these people will require more water, waste treatment systems, food production, energy, transportation systems and manufacturing – all of which require engineering professionals to participate in land planning and to research, study, design, construct and operate new and expanded facilities. This future built environment must be developed while sustaining the natural resources of the world and enhancing the quality of life for all people. Top priority must be placed on sustainable development because of its global importance today.

Over the last few years, the world community has focused on a number of sustainable development issues for which members of the engineering profession can, and must, take a leading role in improving understanding. The following issues are but a few for which part of the solution is technological:

- Climate change is important for us all and the projected changes will bring difficulties to all communities. The evidence is clear that there will be increasingly severe weather events leading to greater incidences and severity of natural disasters. Engineering professionals can assist in mitigating further effects of climate change by developing energy-efficient user products and industrial processes, and by enhancing renewable energy technologies. Engineering professionals can also assist communities to be safer, experience fewer disruptions and lose fewer lives by creating safer, adaptable and resilient buildings and structures.
- Energy production has been raised in profile because of the cost of fuels, environmental impacts and the development of renewable energy sources. Engineering professionals can assist in reducing dependence on high-cost fuels by developing low-energy products and appliances that can be cost effectively run off renewable energy. Engineering professionals can also assist developing countries in securing their energy networks by selecting the most appropriate energy sources and by creating reliable and innovative systems to deliver the energy where it is needed.
- Water scarcity is a high-risk reality in many developed and developing counties. Engineering professionals can assist in providing water security by developing water-efficient and waterless products and processes, and by creating integrated, round-put processes where water is reused and recycled.
- Material waste volumes are increasing in almost every country and threaten to continue to escalate with population growth. Engineering professionals can assist in reducing waste rates by developing

durable, high-value, low-waste products and processes. Engineering professionals can then assist in stabilizing waste through designing products for end of life and by creating integrated, round-put processes where the waste of one component becomes the food of another. Finally, engineering professionals can assist in reversing waste rates by developing innovative products and processes that use and consume existing waste.

- Many of our emissions and material wastes are also toxic and find their way into the air, soil and waterways that support humans and all other organisms on Earth. Engineering professionals can assist in protecting the integrity of the natural environment human health by developing products and processes that use clean energy sources, benign materials and produce benign emissions and wastes. Engineering professionals can also assist developing countries to leapfrog the developed world's last few decades of wasteful and toxic practices and technologies by selecting the most appropriate solutions for their transitioning economies.

It is now recognized that engineering professionals need considerable support in enhancing the practice of engineering to address these issues and to promote sustainable development. Education on sustainable development issues must be given the highest priority. Engineering professionals will be involved in promoting, planning and implementing development in the future and will require the skills to develop and implement sustainable technologies.

This book's contribution to the discussion and theory about sustainable solutions and Whole System Design is an important step to ensure that engineers integrate the theory and practice within their regular design activity. Taking the broader view and the consideration of the widest set of factors into design is now an imperative if the engineering community is to develop its commitment to sustainability. This book is an important contribution to ensuring that the broadest possible gains are achieved from the current interest in life-cycle and ecological costing of products and projects.

The authors, in producing this introductory technical teaching material and these important examples, have provided a publication that can, and must be, widely used in our university and technical training institutions. The way in which the material is presented makes it a valuable reference handbook. The examples highlight the simple application of the theory presented and make the book suitable for self-learning as well as in classroom or tutorial use.

The team at The Natural Edge Project is to be complimented on their preparation of such a valuable resource. Everyone working and studying in this field of engineering should buy it and use it.

Barry J. Grear AO
President, World Federation of Engineering Organizations (WFEO) 2007–2009
Paris, France
October 2008

III

The need for sustainable environmental, social and economic development, with specific reference to such issues as climate change, is one of the major challenges we face both today and into the future. The importance of environmental sustainability is underlined as one of the eight Millennium Development Goals (MDGs) in developing and least-developed countries, and the Intergovernmental Panel on Climate Change (IPCC) has emphasized the importance of technology in climate change mitigation and adaptation.

Despite this, the role of engineering and technology in sustainable social and economic development is often overlooked. At the same time, there is a declining interest and enrolment of young people, especially young women, in engineering. This will have a serious impact on capacity in engineering, and our ability to address the challenges of sustainable social and economic development, poverty reduction and the other MDGs.

The development and application of knowledge in engineering and technology underpins and drives sustainable social and economic development. Engineering and technology are vital in addressing basic human needs, poverty reduction and sustainable development to bridge the 'knowledge divide' and promote international dialogue and cooperation.

What can we do to promote the public understanding of engineering, and the application of engineering in these vital contexts? It appears that the decline of interest and entry of young people into science and engineering is due to the fact that these subjects are often perceived by young people as nerdy, uninteresting and boring; that university courses are difficult and hard work; that jobs in these areas are not well paid; and that science and engineering have a negative environmental impact. There is also evidence that young people turn away from science around the age of ten, that good science education at primary and secondary levels is vital, and that science teaching can turn young people off as well as on to science. There are clear needs to show that science and engineering are inherently interesting and to promote public understanding and perception, to make education and university courses more interesting, with better salary scales (although this is already happening through supply and demand), and to promote science and engineering as part of the solution, rather than part of the problem of sustainable development.

The promotion of public understanding and interest in engineering is facilitated by presenting engineering as part of the problem-solving solution to sustainable development and poverty reduction. University courses can be made more interesting through the transformation of curricula and pedagogy, and more activity, project and problem-based learning, just-in-time approaches and hands-on applications rather than the more formulaic approaches that turn students off. These approaches promote the relevance of engineering, address contemporary concerns and help to link engineering with society in the context of related ethical issues, sustainable development, poverty reduction, and building upon rather than displacing local and indigenous knowledge. The growth of Engineers without Borders and similar groups around the world demonstrates the attractiveness of participating in finding solutions to today's 'real world' problems; the young seem to have a common desire to 'do something' to help those in need.

Science and engineering have changed the world, but are professionally conservative and slow to change – we need innovative examples of schools, colleges and universities around the world that have pioneered activity in such areas as problem-based learning. It is also interesting to look at reform and transformation in other professions – such as medicine, where some of the leading medical schools have changed to a 'patient-based' approach. If the medics can do this when there is no enrolment pressure, then so can engineers. Engineers practice just-in-time techniques in industry; why not in education?

Transformation in engineering education needs to respond to rapid change in knowledge production and application, emphasizing a cognitive problem-solving approach, synthesis, awareness, ethics, social responsibility, experience and practice within national and global contexts. We need to learn how to learn and to emphasize the importance of lifelong and distance learning, continuous professional development, adaptability, flexibility, inter-disciplinarity and multiple career paths.

Such transformation of engineering and engineering education is essential if engineering is to catch and surf the 'seventh wave' of technological revolution – relating to knowledge for sustainable development, climate change mitigation and adaptation, and new modes of learning. This follows the sixth wave of new modes of knowledge generation, dissemination and application, and knowledge and information societies and economies in such areas as information and communications technology (ICT), biotechnology, nanotechnology, new materials, robotics and systems technology, characterized by cross-fertilization and fusion, innovation, the growth of new disciplines and the decline of old disciplines, where new knowledge requires new modes of learning. The fifth wave of technological revolution is based on electronics and computers, the fourth wave on oil, automobiles and mass production, the third wave on steel, heavy engineering and electrification, the second wave on steam power, railways and mechanization, and the first wave on the technological and industrial revolution, and the development of iron and water power.

The main applications challenges relate to how engineering and technology may most effectively be developed, applied and innovated to reduce poverty, promote sustainable development and address climate change mitigation and adaptation. It is apparent that these challenges are linked to a possible solution – many young people and student engineers are keen to address international issues, especially poverty reduction and sustainable development. This is reflected, as mentioned above, by the interest of young people in Engineers without Borders groups around the world and the United Nations Educational, Scientific and Cultural Organization (UNESCO)–Daimler Mondialogo Engineering Award. To promote engineering and attract young people, we need to emphasize these issues in teaching curricula and practice.

In the context of the need for transformation in engineering education to include sustainable development and wider social and ethical issues, the work of the Engineering Sustainable Solutions Programme of The Natural Edge Project, and this publication on *Whole System Design: An Integrated Approach to Sustainable Engineering*, could not be more timely and relevant. It is also important because while the need for whole/holistic and integrated systems approaches in engineering have been recognized and spoken about for some time, there is still a need to share information on what this means in practice, and to share pedagogical approaches and curricula developed in this context. This is particularly important for universities and colleges in developing countries, who face serious constraints regarding human, financial and institutional resources to develop such curricula and learning/teaching methods. It is also timely in view of the United Nations Decade of Education for Sustainable Development, 2005–2014, for which UNESCO is the lead agency.

Engineering is about systems, and so it should be taught. Engineers understand systems, and Nature is the very epitome of a whole system – so it is surprising that engineers have not been more interested in holistic and whole systems approaches in the past. Engineering, however, derives from the 17th-, 18th- and 19th-century knowledge models and 'modern science' of Galileo, Descartes and Bacon, based on reductionism and the objectification and control of Nature. So the rediscovery of holistic thinking is perhaps not surprising and, indeed, overdue, prompted, for example, by the renewed interest in biomimetics that links engineering and technology with natural life structures and systems. This marks a belated return to the biomimetics of Leonardo da Vinci in the 15th and 16th centuries, although this rediscovery has been facilitated by the development of computer science and technology and new materials – one wonders what Leonardo would have done with computer aided design (CAD)/computer aided manufacturing (CAM) and carbon fibre!

This publication is supported by the Department of the Environment, Water, Heritage and the Arts of the Australian government, Engineers Australia and Earthscan, and we look forward to further support of such initiatives, especially now that Australia has signed the Kyoto Protocol. UNESCO supported the production of earlier material on engineering and sustainable development by The Natural Edge Project, and is very happy to be associated with this innovative initiative. I would like to congratulate Peter Stasinopoulos, Michael Smith, Charlie Hargroves and Cheryl Desha on their pioneering activity, and look to continued cooperation with The Natural Edge Project on this area of increasing importance.

Dr Tony Marjoram
Senior Programme Specialist,
United Nations Educational, Scientific and Cultural Organization (UNESCO)
Paris, France
October 2008

IV

This is the challenge:

Around 9 billion people will be living on this Earth in the middle of this century. They will all want to conduct a decent life. They will want a certain minimum standard of material wealth, requiring food, water, shelter and the basic services now taken for granted in our advanced civilizations. However, resources are limited, our climate is vulnerable and changing, and the restorative capacities of ecosystems are declining rapidly. Let us look, for example, at the restrictions related to climate. Consider that greenhouse gas emissions will roughly double by mid century if we continue with business as usual. However, stabilizing our climate requires at least halving greenhouse gas emissions. In addition, many analysts now tell us that we have but precious few decades to do so.

What are some potential solutions?

Renewable energy technologies such as wind and solar power currently provide a small quantity of our total energy requirements and will likely take many years to expand sufficiently. Renewable fuels such as biofuels require large areas of land for crops, and compete with and drive up the price of grain staples. Biofuels are also expensive, even in the European Union where they are more cost competitive since many other technologies require the purchase of permits to emit carbon dioxide.

Nuclear energy is more ecologically controversial, more economically costly and more socially disruptive than biofuels, and the insurance industry refuses to cover the full range of risks. Integrated gasification combined cycle (IGCC) systems combined with carbon capture and storage (CCS) so far look like an expensive dream.

So aren't there any options that meet our energy requirements without emitting excessive greenhouse gases?

The answer may lie in a more radical approach. Why not reinvent technological progress and develop the appropriate changes of behaviour? Imagine a 10kg bucket of water. How much electricity would you need to lift the bucket from sea level to the top of Mount Everest? It may come as a surprise that you would need only one quarter of a kilowatt hour (kWh). Meaning that a kilowatt hour is an amazing powerhouse! But what do we do with one quarter of a kilowatt hour? We power a single 75W incandescent lamp for 3.3 hours. I submit that we can realize far more economic and social benefits than we currently do with each kilowatt hour of energy and, indeed, each kilogram of material and water, each kilometre of transport and each square metre of the Earth's surface.

A fivefold increase in resource productivity, I suggest, will make our ecological and social challenges manageable – that is, a fivefold increase in energy productivity, materials productivity, water productivity, transport productivity and land productivity. Rich countries could stabilize their wealth while reducing their energy and resource consumption by 80 per cent. Poor countries would be encouraged to grow fivefold while stabilizing their demand on resources. In order to achieve these significant improvements, we will require a paradigm shift in productivity. Labour productivity has risen 20-fold since 1850, and now we also require resource productivity to rise. It is important to note here that productivity is more than efficiency. Efficiency is measured in the closed box of a distinct function, such as the kilometres that a car can drive on 1 litre of fuel. Productivity, on the other hand, is measured in a broader perspective of the solution, such as the equivalent transport services that other mobility modes provide per input of a specified resource.

There are, of course, some low hanging fruits, such as efficient lighting, hybrid cars, energy efficient buildings, water purification and waste recycling: combined, they might take care of a factor of two in resource productivity. Achieving a factor of five (80 per cent) increase in resource productivity calls upon our creativity and ability to innovate as we search for new ways of redesigning technologies, processes, infrastructure and systems. Focusing only on optimization at the component level of a system will not deliver the resource productivity needed – optimization at the system level is critical. Systemic improvements go considerably further than isolated component improvements. Synergies between components and cascades of resource use are abundantly available but have to be identified and properly designed in order to deliver the resource productivity needed.

To this end, the team from The Natural Edge Project, led by Charlie Hargroves, offers this book to those wishing to use design to deliver the types of improvements I call for above by taking an integrated approach to sustainable engineering. I was thrilled and impressed reading this volume, which features an integrated approach towards resource productivity and, ultimately, sustainability both at a small and large scale. Each chapter in this book is self-explanatory and self-sufficient, making for easy reading and teaching; but taken as a whole, it is a wonderful contribution to engineering design, as you would expect from a book with this title. Good luck readers, students and teachers!

Professor Ernst Ulrich von Weizsäcker
Co-recipient of the 2008 DBU
German Environmental Award
Lead author of Factor Four *(1995)*
Lead author of Factor Five *(2009)*
Former Chairman, Bundestag Environment Committee
Former President of the Wuppertal Institute for Climate, Environment and Energy
Emmendingen, Germany
October 2008

Acknowledgements

The Natural Edge Project (TNEP) would like to thank the following individuals and groups for making the development of this publication possible. Firstly, a special thank you must go to the authors' families. Peter would like to thank his family and friends for their love and support, especially his family Bill, Georgina, George, Steven and Olivia, and partner Jacquelina. Mike would like to thank his wife Sarah Chapman for her love, support and for sharing a lifelong passion for sustainable engineering. Charlie would like to thank his wife, Stacey, for her patience and love. Cheryl would like to thank her family for their love and support of her commitment to make a difference. The authors would also like to thank Fatima Pinto for her tireless efforts in managing the TNEP office.

TNEP Secretariat – Charlie, Michael, Cheryl, Peter, Stacey Hargroves and Fatima Pinto – would like to thank the Australian Federal Department of the Environment, Water, Heritage and the Arts (DEWR) for funding the development of the publication as part of the 2005/06 and 2006/07 Education for Sustainability Grants Program.

A special thank you must go to Amory Lovins as he was the inspiration for this publication, in particular the starting point for the development of the methodology, and the unique format of the case studies. During our trip to Rocky Mountain Institute in 2004, we asked Amory what a team of young engineers could do to make a difference to our profession and he responded simply that we should contribute to the '*non-violent overthrow of bad engineering*', and the many conversations that followed inspired our team to develop this book.

Thank you to Paul Compston and Benjamin S. Blanchard for taking the time to mentor our team on Systems Design and Systems Engineering. Additional thanks must go to Paul for trialing the book's material in his Systems Design course at The Australian National University. A special thank you goes to Alan Pears for taking the time to share with us his personal experiences and lessons learnt from whole system design projects to inform the development of the methodology on which this book is based.

The Secretariat would also like to thank Barry Grear AO, Benjamin S. Blanchard, Ernst Ulrich von Weizsäcker, and Tony Marjoram for taking the time to mentor our team and contribute forewords for this publication. We would like to thank the following individuals for taking the time to provide peer review and mentoring for this publication:

Al Blake, Royal Melbourne Institute of Technology
Alan Pears, Royal Melbourne Institute of Technology
Angus Simpson, University of Adelaide
Benjamin S. Blanchard, Virginia Polytechnic Institute and State University
Bolle Borkowsky, CDIF Group
Bruce R. Munson, Iowa State University
Chandrakant Patel, Hewlett-Packard
Colin Kestel, University of Adelaide
Dylan Lu, University of Sydney
Janis Birkeland, Queensland University of Technology
Kazem Abhary, University of South Australia
Lee Luong, University of South Australia
Mehdi Toophanpour Rami, University of Adelaide
Nick Edgerton, AMP Capital Sustainability Fund (formerly of the University of Technology Sydney Institute of Sustainable Futures)
Paul Compston, The Australian National University
Philip Bangerter, Hatch
Robert Mierisch, Hydro Tasmania Consulting
Veronica Soebarto, University of Adelaide
Wim Dekkers, Queensland University of Technology

The work was copy-edited by TNEP Professional Editor Stacey Hargroves.

Work on original graphics and enhancements to existing graphics has been carried out by Mr Peter Stasinopoulos, Mrs Renee Stephens and Earthscan.

Author Biographies

The Natural Edge Project (TNEP) is an independent sustainability think-tank based in Australia, which operates as a partnership for education, research and policy development on innovation for sustainable development. TNEP's mission is to contribute to and succinctly communicate leading research, case studies, tools, policy and strategies for achieving sustainable development across government, business and civil society. The team of early career professionals receives mentoring and support from a wide range of experts and leading organizations, in Australia and internationally. Since forming in 2002, TNEP have developed a number of internationally renowned books on sustainable development, which include contributions from colleagues Alan AtKisson, Amory Lovins, Ernst von Weizsäcker, Gro Brundtland, Jeffery Sachs, Jim McNeill, Leo Jensen, R. K. Pachauri and William McDonough.

Peter Stasinopoulos is the Technical Director of The Natural Edge Project. He is a graduate of the University of Adelaide, holding a Bachelor of Mechatronic Engineering with First Class Honours and a Bachelor of Mathematical and Computer Science, and is currently completing a PhD in Systems Design under Dr Paul Compston and Dr Barry Newell at The Australian National University. Since starting with TNEP in 2005, Peter has worked on a variety of projects across TNEP's Education and Industry Consultation portfolios.

Michael Smith is a co-founder and the Research Director of The Natural Edge Project. Michael is also a co-author and co-editor of *The Natural Advantage of Nations* (Earthscan 2005) and co-author of *Cents and Sustainability* (Earthscan 2009). Michael is a graduate of the University of Melbourne, holding a Bachelor of Science with a double major in Chemistry and Mathematics with Honours and has submitted his PhD thesis entitled *Advancing and Resolving the Great Sustainability Debates* under Professor Steve Dovers and Professor Michael Collins at The Australian National University.

Karlson 'Charlie' Hargroves is a co-founder and the Executive Director of The Natural Edge Project. Charlie is also a co-author and co-editor of *The Natural Advantage of Nations* (Earthscan 2005) and co-author of *Cents and Sustainability* (Earthscan 2009). Charlie graduated from the University of Adelaide, holding a Bachelor of Civil and Structural Engineering and is currently completing a PhD in Sustainable Industry Policy under Professor Peter Newman at Curtin University. Prior to co-founding TNEP in 2002, Charlie worked as a design engineer for two years. Charlie spent 12 months on secondment as the CEO of Natural Capitalism Inc, Colorado, and represents the team as an Associate Member of the Club of Rome.

Cheryl Desha is the Education Director of The Natural Edge Project and a lecturer in the School of Engineering at Griffith University. She is a co-author of *The Natural Advantage of Nations* (Earthscan 2005). Cheryl is a graduate of Griffith University, holding a Bachelor of Environmental Engineering with First Class Honours and receiving a University Medal and Environmental Engineering Medal. She is currently completing a PhD in Education for Sustainable Development under Professor David Thiel at Griffith University. Prior to joining TNEP in 2003, Cheryl worked for an international consulting engineering firm for four years. In 2005, she was selected as the Engineers Australia Young Professional Engineer of the Year.

1

A Whole System Approach to Sustainable Design

Educational aim

Chapter 1 explains the importance and relevance of a Whole System Approach to Sustainable Design in addressing the pressing environmental challenges of the 21st century. It introduces the main concepts of a Whole System Approach to Sustainable Design and how it complements 'design for environment' and 'design for sustainability' strategies. It also introduces the need to innovate efficient holistic solutions to reduce our negative impact on the environment and reduce our dependence on fossil fuels. An outline is given of the numerous benefits that Whole System Design brings to business and the nation. These include how Whole System Design can help to achieve sustainable development by enabling the decoupling of economic growth from environmental pressure. The chapter concludes with a summary of the main concepts of Whole System Design that can be used to deliver such solutions. In this book the terms 'Whole System Design', a 'Whole System Approach to Sustainable Design', a 'Whole System Approach to Design' and 'Sustainable Design' are used interchangeably.

Required reading

Environment Australia (2001) *Product Innovation: The Green Advantage: An Introduction to Design for Environment for Australian Business*, Commonwealth of Australia, Canberra, pp1–10, www.environment.gov.au/settlements/industry/finance/publications/producer.html, accessed 5 January 2007

Pears, A. (2004) 'Energy efficiency – Its potential: Some perspectives and experiences', background paper for International Energy Agency Energy Efficiency Workshop, Paris, April 2004, pp1–13

Porter, M. and van der Linde, C. (1995) 'Green and competitive: Ending the stalemate', *Harvard Business Review*, September/October, Boston, MA, pp121–134

Rocky Mountain Institute (1997) 'Tunnelling through the cost barrier', *RMI Newsletter*, Summer 1997, pp1–4, www.rmi.org/images/other/Newsletter/NLRMIsum97.pdf, accessed 5 January 2007

Why does design matter?

As Amory Lovins et al wrote in *Natural Capitalism*:[1]

> By the time the design for most human artefacts is completed but before they have actually been built, about 80–90 per cent of their life-cycle economic and ecological costs have already been made inevitable. In a typical building, efficiency expert Joseph Romm explains, 'Although up-front building and design costs may represent only a fraction of the building's life-cycle costs, when just one per cent of a project's up-front costs are spent, up to 70 per cent of its life-cycle costs may already be committed. When seven per cent of project costs are spent, up to 85 per cent of life-cycle costs have been committed. That first one per cent is critical because, as the design adage has it, "all the really important mistakes are made on the first day".'

Infrastructure, buildings, cars and many appliances all have long design lives, in most cases from 20 to 50 years. The size and duration of infrastructure and building developments, for instance, demand that they should now be much more critically evaluated for efficiency and function than ever before. Australian Ambassador to the United Nations Robert Hill, talking about the new Australian Parliament House, sums up the loss of opportunities from a failure to incorporate environmental considerations into design:[2]

> Across Lake Burley Griffin is one of Australia's most famous houses – Parliament House. Built at considerable cost to the Australian taxpayer, it was officially opened in 1988. Since 1989, efforts have been made to reduce energy consumption in Parliament House, resulting in a 41 per cent reduction in energy use with the flow-on effect of reducing greenhouse gas emissions by more than 20,000 tonnes annually. This has also brought about a saving of more than AU$2 million a year in running costs. But the new wave of environmental thinking would have us question why these measures weren't incorporated in the design of the building in the first place and what other opportunities for energy-saving design features were missed? It's a simple example of how the environment is still considered an add-on option as opposed to being central to the way we do business.

Currently considerable opportunities are being missed at the design phase of projects to significantly reduce negative environmental impacts. There is a great deal of opportunity here for business and government to reduce process costs, and achieve greater competitive advantage through sustainable engineering designs. As Robert Hill also stated:[3]

> Building construction and motor vehicles are two high-profile industry sectors where producers are utilizing Design for Environment (DfE) principles in their product development processes, thereby strategically reducing the environmental impact of a product or service over its entire life-cycle, from manufacture to disposal. Companies that are incorporating DfE are at the forefront of innovative business management in Australia. As the link between business success and environmental protection becomes clearer, visionary companies have the opportunity to improve business practices, to be more competitive in a global economy and to increase their longevity.

The Department of the Environment and Heritage has published an introduction to DfE for Australian businesses, *Product Innovation: The Green Advantage*,[4] which highlights the benefits of pursuing a DfE approach. This is backed up by numerous studies. DfE provides a new way for business to cost-effectively achieve greater efficiencies and competitiveness from product redesign. Harvard Business School Professor Michael Porter, author of *The Competitive Advantage of Nations*, and Claas van der Linde highlight a range of ways that DfE at the early stages of development of a project can both reduce costs and help the environment in their 1995 paper '*Green and Competitive*'.[5]

Some of businesses' most significant costs are capital and inputs such as construction materials, raw materials, energy, water and transportation. It is therefore in businesses' best interests to minimize these costs, and hence the amounts of raw materials and other inputs they need to create their product or provide their service. Business produces either useful products and services or waste, better described as unsaleable production, because the company pays to produce it. How does it assist a business to have plant equipment and labour tied up in generating waste? Table 1.1 below lists the numerous ways companies can profitably reduce waste. Addressing such opportunities therefore gives businesses numerous options to reduce costs and create new product differentiation.

A DfE approach to reducing environmental impacts is one of the best approaches business and government can take to find win-win opportunities to both reduce costs and help the environment. The DfE approach is reminiscent of the 'total quality movement' in business in the 1980s, where many were sceptical at the beginning that re-examining current business and engineering practices would make a difference. Many doubted that win-win opportunities could be found. Today, on the other hand, it is assumed that such win-win opportunities exist if business takes a total quality approach. The Department of the Environment and Heritage publication *Product Innovation: The Green Advantage* showed that many companies are finding win-win ways to reduce costs and improve product differentiation through a DfE approach. Expanding on this concept, companies and government programmes are finding that if a Whole System Design approach is taken, then the cost savings and environmental improvements can be in the order of Factor 4–10 (75–90 per cent).

Table 1.1 *DfE and business competitive advantage*

DfE can Improve Processes and Reduce Costs:	DfE Provides Benefits to Reduce Costs and Create Product Differentiation:
• Greater resource productivity of inputs, energy, water and raw materials to reduce costs; • Material savings from better design; • Increases in process yields and less downtime through designing out waste and designing the plant and process to minimize maintenance and parts; • Better design to ensure that by-products and waste can be converted into valuable products; • Reduced material storage and handling costs through 'just in time' management; • Improved OH&S; and • Improvements in the quality of product or service.	• Higher quality, more consistent products; • Lower product costs (e.g. from material substitution, new improved plant efficiencies etc); • Lower packaging costs; • More efficient resource use by products; • Safer products; • Lower net costs to customers of product disposal; • Higher product resale and scrap value; and • Products that meet new consumer demands for environmental benefits.

Source: Adapted from Porter and van der Linde (1995), p126[6]

This book discusses a Whole System Approach to Sustainable Design. It is important here to discuss the meaning of the term 'Sustainable Design' in this context, where the focus is primarily on technical engineered systems. Sustainable Design refers to the design and development of systems that, *throughout their lifecycle*:

- Consume natural resources (energy, materials and water) within the capacity for them to be regenerated (thus favouring renewable resources), and preferably replace or reuse natural resources;
- Do not release hazardous or polluting substances into the biosphere beyond its assimilative capacity (thus zero release of hazardous persistent and/or bio-accumulative substances), and preferably are benign and restorative;
- Avoid contributing to irreversible adverse impacts on ecosystems (including services and biodiversity), biogeochemical cycles and hydrological cycles, and preferably protect and enrich ecosystems, biogeochemical cycles and hydrological cycles;
- Provide useful and socially accepted services long term, and enrich communities and business by providing multiple benefits; and
- Are cost effective and have a reasonable rate of return on total life-cycle investment, and preferably are immediately profitable.

Currently, not all systems will reflect the above description of a sustainable system. However, almost all systems can be improved towards this end. A number of leading Sustainable Design experts – Bill McDonough,[7] Paul Hawken, Amory Lovins[8], Hunter Lovins,[9] Karl-Henrik Robert,[10] Paul Anastas,[11] Friedrich Schmidt-Bleek,[12] and Sim Van der Ryn[13] – have developed guides to Sustainable Design that are in accord with the criteria outlined above. There are also many other important criteria in developing systems that are 'sustainable' throughout their life-cycle in the traditional sense – in other words, their services are reliable, maintainable, supportable, available and producible.[14]

A Whole System Approach explained

> **A Whole System Approach** is a process through which the interconnections between sub-systems and systems are actively considered, and solutions are sought that address multiple problems via one and the same solution.

In the past engineers have failed to see these large potential energy and resource savings, because they have been encouraged to optimize only parts of the system – be it a pumping system, a car or a building. Engineers have been encouraged to find efficiency improvements in part of a plant or a building, but rarely encouraged to seek to re-optimize the whole system. 'Incremental product refinement' has been traditionally undertaken by isolating one component of the technology and optimizing the performance or efficiency of that one component. Though this method has its merits with the traditional form of manufacturing and management of engineering solutions, it prevents engineers from achieving significant energy and resource efficiency savings. Over the last 20 years, engineers using a Whole System Approach to design has enabled designers to achieve Factor 4–20 (75–95 per cent) efficiency improvements, which in many cases has opened up new more cost–effective ways to reduce our load on the environment. This is because in the past many engineered systems did not take into account the multiple benefits that can be achieved by considering the *whole system*.

For example, as the Rocky Mountain Institute points out, most energy-using technologies are designed in three ways that are intended to produce an optimized design but actually produce suboptimal solutions:

1 Components are optimized in isolation from other components (thus 'pessimizing' the systems of which they are a part).
2 Optimization typically considers single rather than multiple benefits.
3 The optimal sequence of design steps is not usually considered.[15]

Hence the Whole System Approach is now recognized as an important approach to enable the achievement of Sustainable Design. To illustrate this, consider the work of Interface Ltd engineer Jan Schilham in designing an industrial pumping system for a factory in Shanghai in 1997, as made famous largely by Amory Lovins and profiled in *Natural Capitalism*:[16]

> One of its industrial processes required 14 pumps. In optimizing the design, the top Western specialist firm sized the pump motors to total 95 horsepower. But by applying methods learned from Singaporean efficiency expert Eng Lock Lee (and focusing on reducing waste in the form of friction), Jan Schilham cut the design's pumping power to only seven horsepower – a 92 per cent or 12-fold energy saving – while reducing its capital cost and improving its performance in every respect.

Schilham did this in two simple ways. First, he revisited pipe width. The friction in pipes decreases rapidly (nearly to the fifth power) as the diameter increases. He found that the existing pipe arrangement wasn't taking advantage of this mathematical relationship, and so he designed the system to use short, fat pipes instead of long, thin ones. Second, he adjusted the system to minimize bends in pipes (to further reduce friction). This Whole System Approach created a 12-fold reduction in the energy required to pump the fluids through the pipe system, resulting in the big reduction in motor size, and subsequent energy and cost savings. Why is this significant? As Amory Lovins writes:

> Pumping is the biggest use of the motors, and motors use 3/5 of all the electricity, so saving one unit of friction in the pipe save 10 units of fuel. Because of the large amount of losses of electricity in its transmission from the power plant to the end use, saving one unit of energy in the pump/pipe system saves upwards of ten units of fuel at the power plant.[17]

A Whole System Approach to Sustainable Design allows multiple benefits to be achieved in the design of air-handling equipment, clean-rooms, lighting, drivepower systems, chillers, insulation, heat-exchanging and other technical systems in a wide range of sizes, programmes and climates. Such designs commonly yield energy savings of 50–90 per cent. However, only a tiny fraction of design professionals routinely apply a Whole System Approach to Sustainable Design. Most design projects deal with only some elements of an energy/materials-consuming system and do not take into account the whole system. This is the main reason why they fail to capture the full savings potential. A Whole System Approach to Sustainable Design is increasingly being seen as the key strategy to achieving cost-effective ways to reduce negative environmental impacts.

This was one of the main conclusions of the five-year Australian Federal Government Energy Efficiency Best Practice (EEBP) programme run by the Department of Industry, Tourism and Resources (DITR).[18] The team involved found that through a 'whole-of-system' approach they could achieve 30–60 per cent energy efficiency gains across a wide range of industries, from bakeries to supermarkets, mines, breweries, wineries and dairies, to name but a few. The programme explicitly recommends that project teams take a whole-of-system approach to understanding the complex challenges and identifying energy-efficiency opportunities.[19] The programme considered a number of industry applications, including motor systems that are used in almost every industry. It found that electric motors are used to provide motive power for a vast range of end uses, with crushers, grinders, mixers, fans, pumps, material conveyors, air-compressors and refrigeration compressors together accounting for 81 per cent of industrial motive power. The programme pointed out that with a whole-of-system approach to optimizing industrial motor-driven applications, coupled with best practice motor management, electricity savings of 30–60 per cent can be realized.

For example, consider an electric motor driving a pump that circulates a liquid around an industrial site.[20] This system comprises:

- An electric motor (sizing and efficiency rating);
- Motor controls (switching, speed or torque control);
- Motor drive system (belts, gearboxes, etc);
- Pump;
- Pipework; and
- Demand for the fluid (or in many cases the heat or 'coolth' it carries).

The efficiencies of these elements interact in complex ways. However, consider a simplistic situation, where the overall efficiency of the motor is improved by 10 per cent (by a combination of appropriate sizing and selection of a high-efficiency model). The efficiencies of these elements interact in complex ways. However, consider a simplistic situation of a motor system with six components in series. If the efficiency of every component is improved by 10 per cent (by a combination of appropriate sizing and selection of a high-efficiency model), then the overall level of energy use is $0.9 \times 0.9 \times 0.9 \times 0.9 \times 0.9 \times 0.9 = 0.53$. That is 47 per cent savings are achieved. This is why taking the Whole System Approach to Design is yielding over 50 per cent improvements previously ignored in resource productivity, with corresponding reductions in negative environmental impacts. If the most efficient component is chosen for each part of a motor system (even if the difference in efficiency is not significant for the individual components), the overall efficiency of the whole system is about 7 times greater (see Table 1.2).

Table 1.2 *Comparison of the best and the worst efficiency motoring systems*

System Component	Best Efficiency	Worst Efficiency
Electrical wiring	0.98	0.9
Motor	0.92	0.75
Drive (e.g. gearbox or belt)	1.0	0.7
Pump	0.85	0.4
Pipes	0.9	0.5
Process demand	Can vary enormously but assumed constant for this example.	
Overall efficiency of system	0.69	0.095

Source: Pears, A. (2004)[21]

Whole System Design – a rediscovery of good Victorian engineering

During the 20th century, engineering became more and more specialized as scientific and technological knowledge increased exponentially, so much so that now in the 21st century engineers are no longer trained across fields of engineering as they were before and thus no longer keep up with the latest breakthroughs in every field. As a result, opportunities are often missed to optimize the whole system, as the engineer only knows their field in detail and has little interaction with other designers on the project.

A classic example of this is industrial pressurized filtration, which is responsible for over one-third of all the energy used in filtration globally. For the last 80 years most have assumed that these industrial pressurized filters had been designed optimally. However, closer inspection by Professors White, Bogar, Healy and Scales at the University of Melbourne revealed that they had in fact not yet been optimized. The design had been developed 80 years ago by a mechanical engineer who had designed a system which, when given very concentrated suspensions to filter, simply pushed harder rather than adjusting the chemistry of the suspension to make it easier to push through, as the research team from the University of Melbourne have now done. In this case the engineer did not have the training in chemistry, or consult a chemist, to see possibilities to improve the design of the whole system. This clearly demonstrates the benefit of engineers working together across disciplines to examine and optimize engineering systems by pooling their collective knowledge. Most engineering firms have this capacity.

Another factor in why components of the engineering project are optimized in isolation rather than as part of a system is because today large engineering projects are highly complex. Hence the engineer managing the project inevitably has to break up the project into components which are then worked on by individual engineers and designers. Therefore often when undertaking the components of such projects, the individual engineer is not responsible for the whole project and has little choice but to focus on optimizing smaller components of the system, and hence missing those opportunities achievable only through a Whole System Approach to design. But this can be avoided and significant time and money can be saved if extra time is taken at the planning stage of the process to consider Whole System Design opportunities and unleash the creativity of the designers through multidisciplinary design processes such as design charrettes.

Engineers thrive on challenges, and the recently developed field of engineering called 'Systems Engineering' has evolved to address the need on complex engineering projects for an engineer to ensure that all the parts of the project relate and fit. A systems engineer needs to use a Whole System Approach to design and communicate the opportunities effectively to the other engineers involved with developing components of an engineering project. Best practice in Systems Engineering still performs reductionist analyses of engineering challenges, but without losing sight of how one component of the system interacts with and affects all other components of the system or the system's behaviour and characteristics as a whole. As engineers seek to collaborate across the different fields of engineering once more, any Whole System Approach to design involving multidisciplinary engineering teams becomes a rediscovery of the rich heritage of 'Victorian' engineering.

Engineering has a rich tradition of valuing and practising a Whole System Approach to design and optimization. The first industrial revolution, as we know it today, would not have been possible if engineer James Watt had not practised a Whole System Approach to design optimization to achieve major resource productivity gains on the steam engine in 1769. The first industrial revolution was only possible because of the significant improvement in the conversion efficiency of the steam engine[22] thus achieved.[23] Watt realized that the machine was extremely inefficient. Though the jet of water condensed the steam in the cylinder very quickly, it had the undesirable effect of cooling the cylinder down, resulting in premature condensation on the next stroke. In effect the cylinder had to perform two contradictory functions at once: it had to be boiling hot in order to prevent the steam from condensing too early but also had to be cold in order to condense the steam at just the right time.

Watt redesigned the engine by adding a separate condenser, allowing him to keep one cylinder hot by jacketing it in water supplied by the boiler. This cylinder ensured that the water was turned into steam and then another condenser was kept at the right temperature to ensure the steam would condense at just the right time. The result was an immensely more powerful machine than the Newcomen 'steam' engine, the original steam engine.

Watt's initial successful Whole System Design was followed by further remarkable improvements of his own making. The most important of these was the sun-and-planet gearing system, which translated the engine's reciprocating motion into rotary motion. In simple terms, the new machine could be used to drive other machines. Watt alone had used a whole system optimization of the design to turn a steam pump into a machine that had vastly improved resource productivity and applicability.

The need for sustainable Whole System Design

Whole System Design provides ways to both improve conversion efficiency and resource productivity and reduce costs. James Watt showed this over 200 years ago. But in the 21st century it needs to go further. We need to seek to be restorative of the planet rather than destructive, and thus Whole System Design needs to design for sustainability.[24] In other words *we need a Whole System Approach to Sustainable Design*. In the context of the loss of natural capital and the loss of resilience of many of the world's ecosystems, development must be redesigned not to simply harm the environment less, but rather to be truly restorative of nature and ecosystems, and society and communities. This involves the complete reversal of the negative impacts of existing patterns of land use and development, improving human and environmental health, and increasing natural capital (increasing renewable resources, biodiversity, ecosystem services and natural habitat).

To achieve sustainability, we must transform our design and construction processes well beyond what many today see as 'best practice', which merely aims to reduce adverse impacts relative to conventional development in an 'end-of-pipe' manner. Many of what are currently regarded as 'ecological' design goals, concepts, methods and tools are not adequately geared towards the systems design thinking and creativity required to meet this challenge. An entirely new form of design for development is required, of which a Whole System Approach to Sustainable Design, as outlined in this book, provides many of the keys:

> To use an analogy; in the healthcare fields we have moved (conceptually) from (a) alleviating symptoms to (b) curing illness, (c) preventing disease and (d) improving health. Development control is still largely at the first stage – mitigating impacts (in other words alleviating symptoms). Restorative Whole System Approaches to Sustainable Design instead seek to reverse impacts, eliminate externalities and increase natural capital by supporting the biophysical functions provided for by nature to restore the health of the soil, air, water, biota and ecosystems.[25]

Taking a Whole System Approach to Sustainable Design is not simply about reducing harm, but about restoring the environment. It is also about not just ensuring that future generations can meet their needs. A Whole System Approach to Sustainable Design is about designing systems which create a greater array of choices and options for future generations.

One of the leading proponents of Sustainable Design, Bill McDonough tells the following story to illustrate the benefits of a restorative perspective to design. This case study is given in full to give a sense of the potential of design for sustainability:[26]

> In 1993, we helped to conceive and create a compostable upholstery fabric, a biological nutrient. We were initially asked by Design Tex to create an aesthetically unique fabric that was also ecologically intelligent, although the client did not quite know at that point what this would (tangibly) mean. The challenge helped to clarify, both for us and for the company we were working with, the difference between superficial responses such as recycling and reduction and the more significant changes required by the Next Industrial Revolution (and Whole System Design). For example, when the company first sought to meet our desire for an environmentally safe fabric, it presented what it thought was a wholesome option: cotton, which is natural, combined with PET (polyethylene terephthalate) fibres from recycled beverage bottles. Since the proposed hybrid could be described with two important eco-buzzwords, 'natural' and 'recycled', it appeared to be environmentally ideal. The materials were readily available, market-tested, durable and cheap. But when the project team looked carefully at what the manifestations of such a hybrid might be in the long run, we discovered some disturbing facts. When a person sits in an office chair and shifts around, the fabric beneath him or her abrades; tiny particles of it are inhaled or swallowed by the user and other people nearby. PET was not designed to be inhaled. Furthermore, PET would prevent the proposed hybrid from going back into the soil safely, and the cotton would prevent it from re-entering an industrial cycle. The hybrid would still add junk to landfills, and it might also be dangerous.
>
> The team decided to design a fabric so safe that one could literally eat it. The European textile mill chosen to produce the fabric was quite 'clean' environmentally, and yet it had an interesting problem: although the mill's director had been diligent about reducing levels of dangerous emissions, government regulators had recently defined the trimmings of his fabric as hazardous waste. We sought a different end for our trimmings: mulch for the local garden club. When removed from the frame after the chair's useful life and tossed onto the ground to mingle with sun, water and hungry micro-organisms, both the fabric and its trimmings would decompose naturally. The team decided on a mixture of safe, pesticide-free plant and animal fibres for the fabric (ramie and wool) and began working on perhaps the most difficult aspect: the finishes, dyes and other processing chemicals. If the fabric was to go back into the soil safely, it had to be free of mutagens, carcinogens, heavy metals, endocrine disrupters, persistent toxic substances and bio-accumulative substances.
>
> Sixty chemical companies were approached about joining the project, and all declined, uncomfortable with the idea of exposing their chemistry to the kind of scrutiny necessary. Finally one European company, Ciba-Geigy, agreed to join. With that company's help the project team considered more than 8000 chemicals used in the textile industry and eliminated 7962. The fabric – in fact, an entire line of fabrics – was created using only 38

chemicals. The resulting fabric has garnered gold medals and design awards and has proved to be tremendously successful in the marketplace. The non-toxic fabric, Climatex(R)Lifecycle(TM), is so safe that its trimmings can indeed be used as mulch by local garden clubs.

The director of the mill told a surprising story after the fabrics were in production. When regulators came by to test the effluent, they thought their instruments were broken. After testing the influent as well, they realized that the equipment was fine – the water coming out of the factory was as clean as the water going in. The manufacturing process itself was filtering the water. The new design not only bypassed the traditional three-R responses to environmental problems, but also eliminated the need for regulation.

Benefits to business of a Whole System Approach to Sustainable Design

Product improvements and increased competitive advantage

A Whole System Approach to Sustainable Design can help designers to help businesses develop new business opportunities through developing 'greener' products. Such an approach prompts the designer to re-examine existing systems to design totally new ways to meet people's needs, design completely new products, or simply redesign and significantly improve old products. These new product improvements can create new business opportunities, markets and new competitive advantages for a company.

This is being understood by major companies. For instance in May 2005, General Electric (GE), one of the world's biggest companies, with revenues of US$152 billion in 2004, announced 'Ecomagination', a major new business driver expected to double revenues from greener products to US$20 billion by 2010. This initiative will see GE double its research and development in eco-friendly technologies to US$1.5 billion by 2010, and improve energy efficiency by 30 per cent by 2012. In May 2006, the company reported revenues of US$10.1 billion from its energy-efficient and environmentally advanced products and services, up from US$6.2 billion in 2004, with orders nearly doubling to US$17 billion.

Examples of how a Whole System Approach can lead to big advances are now very common:

- Whole System Design improvements mean that refrigerators today use significantly less energy than those built in the early 1980s. In Australia the average refrigerator being purchased is 50 per cent more efficient than the ones bought in the early 1980s. But a Whole System Approach to Sustainable Design motivates the designer to see if this could still be improved. As Chapter 5 will show, the latest innovations in materials science from Europe mean that there are now better insulating materials available that will allow the next generation of refrigerators to be still more energy efficient.
- A Whole System Approach to Sustainable Design involves setting a high stretch goal of seeking to design a system as sustainably and cost effectively as possible. The laptop computer is a classic case study, because it shows what happens when you give engineers a stretch goal. In this case the stretch goal was that computer companies needed laptops to be 80 per cent more efficient than desktop computers so that the computer could run off a battery. With this stretch goal the engineers delivered a solution through Whole System Design.
- The built environment is another major area where many are now taking a Whole System Approach to Sustainable Design. In Melbourne, Australia, the 60L Green Building demonstrated what is possible through retrofitting old buildings with a Whole System Design Approach. This commercial building now uses over 65 per cent less energy and over 90 per cent less water than a conventional commercial building. It features many innovations, using the latest in stylish office amenities completely made from recycled materials.
- Whole System Approaches to Design also can help metal processing and industrial processes. Developed in Australia, Ausmelt was a totally new smelting process for base metals that increased the capacity of metal producers to repeatedly recycle the planet's finite mineral resources. The technology has since been further developed to reprocess toxic wastes such as the cyanide- and fluorine-contaminated pot-lining from aluminium smelters. The Sirosmelt, Ausmelt and Isasmelt technologies have become the system of choice as smelting companies slowly modernize internationally.

Table 1.3 *Case studies of a Whole System Approach to Sustainable Design (as outlined in Chapters 6–10)*

Case Study	Summary
Industrial Pumping Systems	A Whole System Approach to the redesign of a single-pipe, single-pump system focused on a) reconfiguring the layout for lower head loss and b) considering the effect of many combinations of pipe diameter and pump power on life-cycle cost. The WSD system uses 88% less power and has a 79% lower 50-year life-cycle cost than the conventional system.
Passenger Vehicles	A Whole System Approach to the redesign of a passenger vehicle focused on reducing mass by 52% and reducing drag by 55%, which then reduces rolling resistance by 65% and makes a fuel cell propulsion system cost effective. The WSD vehicle is also almost fully recyclable, generates zero operative emissions and has a 95% better fuel-mass consumption per kilometre than the equivalent conventional vehicle.
Electronics and Computer Systems	A Whole System Approach to the redesign of a computer server focused on using the right sized energy-efficient components, which then reduced the heat generated. The WSD server has 60% less mass and uses 84% less power than the equivalent server, which would reduce cooling load in a data centre by 63%.
Temperature Control of Buildings	A Whole System Approach to the redesign of a simple house focused on a) optimizing the building orientation, b) optimizing glazing and shading, and c) using more energy-efficient electrical appliances and lamps. While the WSD house has a AU$3000 greater capital cost than the conventional house, its 29% lower cooling load will reduce energy costs by AU$15,000 over 30 years.
Domestic Water Systems	A Whole System Approach to the redesign of a domestic onsite water system focused on a) using water-efficient appliances in the house and b) optimizing the onsite wastewater treatment subsystem, which then reduces the capacity and cost of the subsurface drip irrigation subsystem, and reduces operating and maintenance costs. The WSD system uses 57% less water and has a 29% lower 20-year life-cycle cost than the conventional system.

Improving competitive advantage through reduced costs

A Whole System Approach to Sustainable Design helps companies move away from end-of-pipe approaches to pollution control, towards designing out waste in the first place and improving eco-efficiency and resource productivity.

Companies are starting to realize that resource inefficiencies in their businesses are often indicators of much greater waste occurring in areas from product design to overall plant design and operation. Professor Michael Porter, internationally renowned expert in business competitiveness, summarizes the key insight that many are still failing to realize, as he and Claas Van der Linde write:[27]

> Environmental improvement efforts have ... focused on pollution control through better identification, processing and disposal of discharges or waste – costly approaches. In recent years, more advanced companies and regulators have embraced the concept of pollution prevention, sometimes called source reduction, which uses such methods as material substitution and closed-loop processes to limit pollution before it occurs. Although pollution prevention is an important step in the right direction, ultimately companies must learn to frame environmental improvement in terms of resource productivity.
>
> Today managers and regulators focus on the actual costs of eliminating or treating pollution. They must shift their attention to include the opportunity costs of pollution – wasted resources, wasted effort and diminished product value to the customer. At the level of resource productivity, environmental improvement and competitiveness come together. This new view of pollution as resource inefficiency evokes the quality revolution of the 1980s and its most powerful lessons. Today many businesspeople have little trouble grasping the idea that innovation can improve quality while actually lowering cost.
>
> But as recently as 15 years ago, managers believed there was a fixed trade-off. Improving quality was expensive because it could be achieved only through inspection and rework of the 'inevitable' defects that came off the line. What lay behind the old view was the assumption that

both product design and production processes were fixed. As managers have rethought the quality issue, however, they have abandoned that old mindset. Viewing defects as a sign of inefficient product and process design – not as an inevitable by-product of manufacturing – was a breakthrough. Companies now strive to build quality into the entire process. The new mindset unleashed the power of innovation to relax or eliminate what companies had previously accepted as fixed trade-offs.

Improved productivity

A Whole System Approach to Sustainable Design can encourage new approaches and innovations that can improve businesses' resource productivity significantly. If industry simply tinkers with the way modes of production currently meet consumer demand, then the productivity gains will be small, but larger resource productivity gains, achieved wisely through Design for Environment and whole system redesign strategies, can help businesses achieve higher productivity gains than usual. Large resource productivity gains can lead to significant total productivity improvements that, to date, have been largely ignored by business due to relatively low energy and water prices and the relatively low costs of landfill. A Whole System Approach to Sustainable Design is a strategy for achieving large resource productivity gains as cost effectively as possible. Numerous business case studies now have been reported to prove this in internationally bestselling books such as *Factor 4: Doubling Wealth, Halving Resource Use* and *Natural Capitalism*. Amory Lovins, a co-author of these books, has been working in recent years with Wal-Mart, the world's largest retailing company. In October 2005, Wal-Mart announced a US$500 million climate change commitment, including initiatives to:

- Reduce greenhouse gas emissions by 20 per cent in seven years; and
- Increase truck fleet fuel efficiency by 25 per cent in three years and double it in ten through a Whole System Redesign of its trucking fleets (to reduce, for instance, their air-resistance).

With the savings from greater energy efficiency, Wal-Mart has also committed to operating on 100 per cent renewable energy.

The Australian Department of Industry, Tourism and Resources (DITR) energy efficiency programme has shown that a Whole System Approach provides a way to achieve large resource efficiency savings while reducing costs to business (see Table 1.4). One area where design can often help businesses save money is by looking at equipment – is it optimized for the job it was intended to do? For instance, most air-conditioners are currently optimized for the most extreme of weather conditions, rather than being optimized for the conditions in a building in which they are required to run most of the time. Another question not often asked is whether there are more systems running than needed. When undertaking a whole system analysis of an industrial plant, office building or factory, it is often found that energy consumption far exceeds the levels expected on the basis of computer simulation. In most systems, from household appliances to office buildings to industrial sites, the nature of energy use can be characterized as shown in Figure 1.1.

In practice, most plant and equipment has surprisingly high fixed energy overheads, because engineers have not checked that what is switched on is only what absolutely needs to be running. In an ideal process, no energy is used when the system is not doing anything useful. The gradient of the graph should reflect the ideal amount of energy used to run the process, but the gradient of the typical process is steeper than the ideal graph, reflecting the inefficiencies within the process (Figure 1.1). Systems ranging from large industrial plants to retail stores to homes show similar characteristics. Why is this happening? It is occurring because, whether the system is an industrial plant or a home, there is very limited measurement and monitoring of energy and resource use at the process level.

Further, rarely are there properly specified benchmarks against which performance can be evaluated. So often plant operators do not know what is possible. An effective strategy looks at both the fixed energy overheads and the system's marginal efficiency. Often only one or the other is addressed. The message here is that energy-consuming systems are not simple. Ideally, they should be modelled under a range of realistic operating conditions, so that appropriate priorities for savings measures can be set and reasonable estimates of energy savings from each measure can be made.

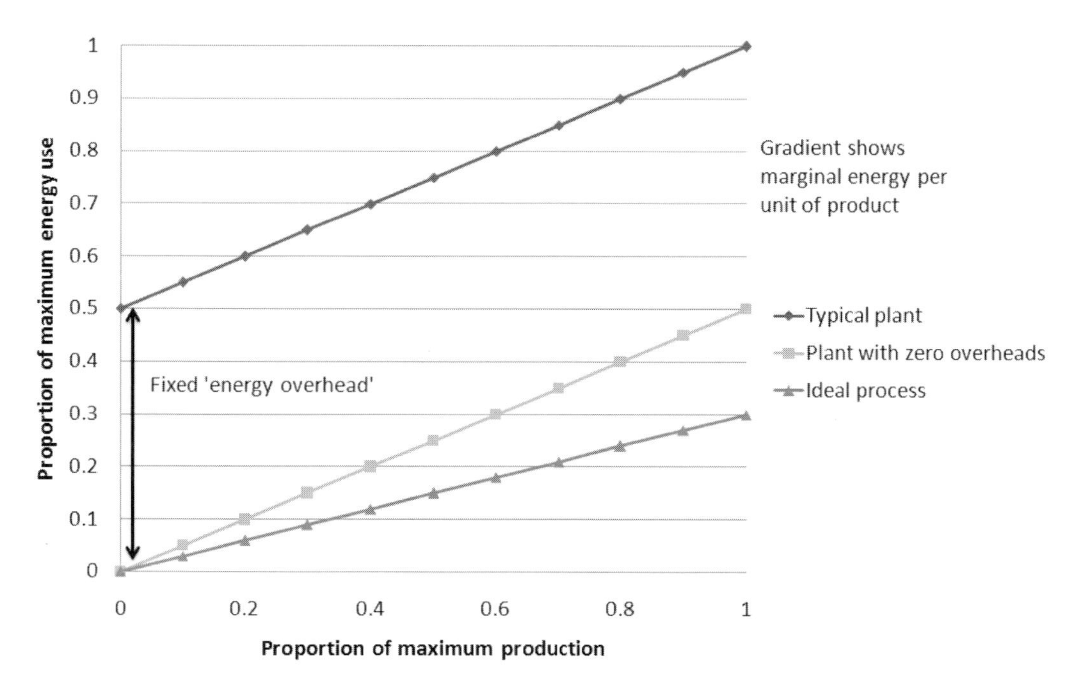

Source: Pears (2004)[28]

Figure 1.1 *Energy use of a typical production system compared with one with zero energy overheads and the ideal process*

Table 1.4 *Sample of the Big Energy Projects (BEP) scheme and Best Practice People and Processes (BPPP) modules under the Energy Efficiency Best Practice government programme*

Site	Core Business	Elements of the programme
Barrett Burston Malting, Geelong, Victoria (and across sites nationally)	Malt Manufacture	BEP (new plant with focus on heating/cooling). BPPP modules: refrigeration compressed air and BEP outcomes workshop.
Savings across six sites in the year to December 2001 yielded an improved energy consumption of around 50,000Gj of combined gas and electricity savings, while maintaining product quality. Total operational costs bettered the budget by 12%, with savings in excess of 20% in one malt house. The improved trend is being continued to this day in all six plants. Significant savings have been identified for the Geelong site and future greenfield sites with the potential to reduce greenhouse gas emissions by 43%.		
Amcor Packaging, Thomastown, Victoria	Bottle Closure Manufacturing	BPPP module: energy management team.
In the first phase, a 'changeover' project was identified by the team, resulting in a productivity increase with a sales value of AU$330,000 annually.		
Amcor Packaging, Dandenong, Victoria	Aluminium Can Manufacturing	BPPP module: energy management.
Efficiency of one gas-fired oven has been improved by 25%, with a saving of 4Gj per hour as well as reliability and productivity benefits. A power factor correction project has been identified that will yield savings of AU$17,000 per year. A compressed-air optimization project has identified savings of AU$46,000 per year.		
Bakers Delight Mascot, Sydney	Bakery	BEP: designed a showcase bakery.
The project achieved 32% savings in annual energy costs and 48% reduction in greenhouse emissions per year compared to a standard Bakers Delight bakery. The project also led to improvements in waste minimization, water conservation and purchasing energy from renewable sources.		

Source: Australian Government's Department of Industry, Tourism and Resources, cited in Hargroves and Smith (2005), p154[29]

Improved decision-making and problem-solving

Initiatives using a Whole System Approach encourage an organization to reconsider outdated processes and assumptions, and can create simultaneous improvements in resource productivity and economic performance. A new approach for engineers to encourage them to re-examine the assumptions underlying long-established manufacturing processes may lead firms to discover opportunities for simultaneously reducing costs and pollutant emissions. Many participants in the US voluntary challenge programmes, such as 33/50 and Green Lights, reported that the programmes forced them to re-examine their decision-making methods. In Australia the Department of Industry, Tourism and Resources (DITR) Energy Efficiency Best Practice programme found that time and again companies can benefit from re-examining assumptions. The sorts of things engineers[30] have found, using a whole system analysis of existing systems, included:

- Large boiler feed water tanks that were uninsulated but sitting in the open air at 75°C. Why? Staff had noted that when the plant wasn't running, the temperature of the water in these tanks fell quite slowly: it was therefore inferred that heat loss was not great. In reality this outcome was due to the very high thermal capacity of a large volume of water, and actual heat losses were hundreds of watts per square metre of surface area, and even more when it was windy or raining.
- Many plants that operated for 50–80 hours per week had large boiler systems or refrigeration systems that could not be shut down and restarted reliably or quickly, so there was massive standby energy waste because they ran more or less continuously.
- Some facilities, for example wineries, had large amounts of high-capital-cost equipment that was fully utilized for very short periods of time: load management strategies offer both capital and energy savings.
- Thermal bridging and air leakage were often major contributors to energy losses and can be easily overcome. For example, a bakery oven evaluated based on actual standby energy consumption had an effective average thermal resistance of R0.22. This compares with a typical insulation value in a house ceiling of R3. The unnecessary 8 kW of heat loss from this oven was a major contributor to the discomfort of staff in the kitchen. In turn, the uncomfortable working conditions are a key factor affecting the difficulty of attracting staff to this industry. The business is actually paying for the energy that undermines its ability to employ good people.

Participants in the DITR energy-efficiency programmes found that taking a Whole System Approach helped their consultant engineers find new ways to address and solve long-standing problems:

> The specialists participating in the workshop were able to consider the malting process from a completely fresh angle, generating a host of valuable creative ideas for future plant designs and many solutions for retrofitting existing plants. ... I heard more innovative ideas about how we can improve our process during this workshop than I've heard in the last 30 years. (Grant Powell, Vice President of Production, Barrett Burston Maltings)

> We found the DITR's Energy Efficiency Best Practice programme to be particularly valuable as a means to incorporate a wide range of external points of view. The specialists involved were able to look at our refrigeration issues without the constraints of having worked in the brewing industry previously. (Phil Browne, Manager Infrastructure and Utilities Capability, CUB)

Benefits to governments

Assist the decoupling of economic growth from environmental pressures

As Yukiko Fukasaku wrote for the OECD in 1999:[31]

> It used to be taken for granted that economic growth entailed parallel growth in resource consumption, and to a certain extent environmental degradation. However, the experience of the last decades indicates that economic growth and resource consumption and environmental degradation can be decoupled to a considerable extent.

> The path towards sustainable development entails accelerating this decoupling process[32] ... in other words transforming what we produce and how we produce it.

The scientific results of the 2005 United Nations Millennium Ecosystem Assessment show that it is vital that all nations achieve rapid decoupling of economic growth from environmental pressures.[33] Many nations, such as The Netherlands (see Figure 1.2), Sweden and the UK, are achieving significant decoupling of economic growth from several environmental pressures, showing that it is possible through eco-efficiencies and Whole System Design to achieve decoupling.[34]

In 2001 the Australian Government committed to the goal of decoupling economic growth from environmental pressures through the then Federal Environment Minister Robert Hill's[36] active participation in, and support for, the 2001–2011 OECD Environmental Strategy, which included 'Achieving Decoupling of Economic Growth from Environmental Pressure' as the second of five key objectives.

Profitable reductions in greenhouse gas emissions

The world's largest economic powers – countries and companies – now acknowledge that greenhouse gas emissions will need to be drastically reduced over the next 30–50 years to avert catastrophic environmental damage leading to significant social and economic damages, as indicated by the IPCC's fourth Assessment (2007). The President of the US, George W. Bush, stated in 2005 that:

> I recognize that the surface of the Earth is warmer and that an increase in greenhouse gases caused by humans is contributing to the problem. (George W. Bush, quoted in *The Washington Post*)[37]

At the 2005 World Economic Forum, CEOs from the world's biggest companies agreed: 'The greatest challenge facing the world in the 21st century – and the issue where business could most effectively adopt a leadership role – is climate change'.[38]

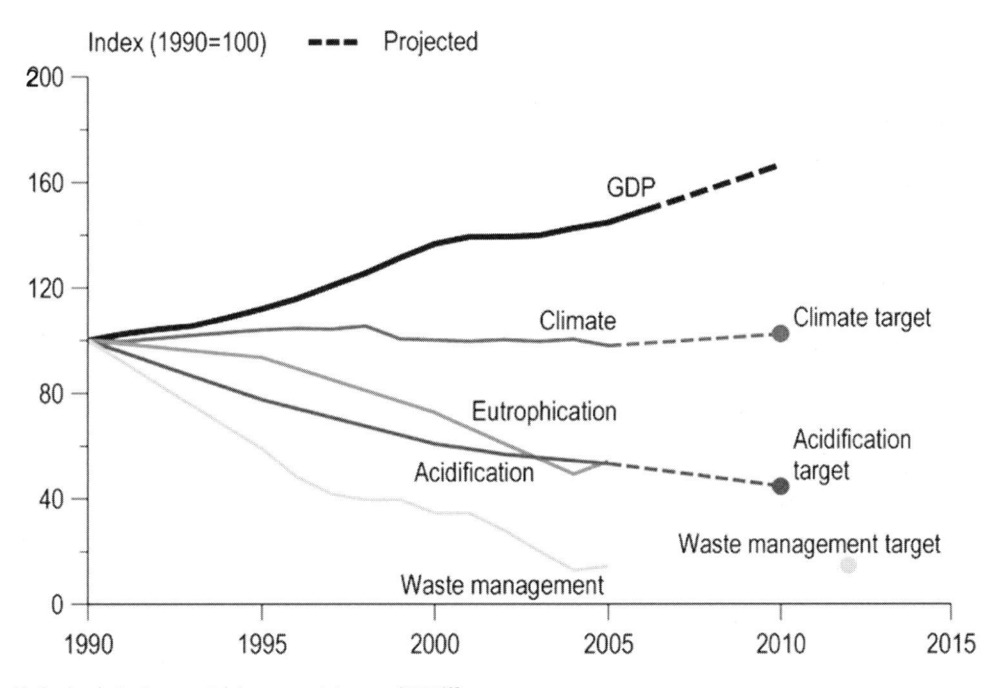

Source: The Netherlands Environmental Assessment Agency (2007)[35]

Figure 1.2 *Comparing The Netherlands' economic growth and reduction of environmental impacts*

While Australia is well on track to achieving its Kyoto target,[39] it is widely acknowledged that this is but a small step in a long journey of greenhouse gas reduction for our country. The Intergovernmental Panel on Climate Change (IPCC) suggest that stabilizing greenhouse gas concentrations at double the pre-industrial levels will require deep cuts in annual global emissions of 60 per cent or more.[40] *The Sydney Morning Herald* reported in 2004 that Australia's Chief Scientist Robin Batterham suggests that an 80 per cent reduction is required in Australia's CO_2 emissions by the end of the 21st century.[41]

A Whole System Approach to Sustainable Design will be a crucial tool to enable the achievement of such large greenhouse gas reductions. As shown above, already there are numerous Whole System Design innovations – pipe and pump systems, motor systems, hybrid cars, laptop computers, and green buildings – which achieve at least 50 per cent energy-efficiency savings. Numerous further case studies of a Whole System Approach to Sustainable Design are outlined in Chapters 4 and 5 showing that 30–80 per cent energy-efficiency savings can be achieved, thus making low-carbon-energy-supply options economically viable. The technical Whole System Design 'worked examples' in Chapters 6–9 – industrial pumping systems, passenger vehicles, electronics and computer systems, and temperature control of buildings – are all examples of how Whole System Design can reduce energy usage and greenhouse gas emissions.

Reducing oil dependency

Reducing our use and dependence on fossil fuels such as oil is not only necessary for reasons associated with global warming – there is also an economic imperative. Whenever oil prices have risen significantly in the past, this has hurt economies in two ways. First, rising oil prices are inflationary and reduce consumer spending in other parts of the economy. US President George W. Bush committed the US to reducing oil dependency by 75 per cent by 2025; he outlined this in his 2006 State of the Union Address, stating:

> And here we have a serious problem: America is addicted to oil, which is often imported from unstable parts of the world. The best way to break this addiction is through technology. Since 2001, we have spent nearly $10 billion to develop cleaner, cheaper and more reliable alternative energy sources – and we are on the threshold of incredible advances. ... This and other new technologies will help us reach another great goal: to replace more than 75 per cent of our oil imports from the Middle East by 2025. By applying the talent and technology of America, this country can dramatically improve our environment, move beyond a petroleum-based economy and make our dependence on Middle Eastern oil a thing of the past.[42]

Modern economies' transportation needs are remarkably dependant on oil. Without new discoveries, Australia's domestic oil reserves are forecast by the Australian Petroleum Association to run out by 2030. Overall oil production has now peaked in over 60 countries (for example, in the US, the rate of oil production peaked in 1972). Increasingly, experts believe that oil production will peak anytime between 2010 and 2030, as shown in Figure 1.3 below. The combination of the approaching oil production peak and increasing oil demand has led to oil prices rising quite rapidly since late 2003 and even more rapidly since early 2007.[43]

Two US Government reports have given serious warnings on this issue and recommended early action. The US Department of Energy's Office of Naval Petroleum and Oil Shale Reserves released a report in 2004 which outlined that, with oil, 'A serious supply-demand discontinuity could lead to worldwide economic crisis.' The authors of this report argue for an emergency plan to keep US oil supplies strong and ensure that the US Naval Fleet can stay afloat.[44] And, in 2005, Robert Hirsh[45] of the Science Applications International Corporation (SAIC) released a report commissioned by the US Department of Energy titled *Peaking of World Oil Production: Impacts, Mitigation and Risk Management*.[46] It delivered a blunt message: that the world has, at most, 20–25 years before world oil production peaks. It argues that it will take economies over 20 years to adapt to a world of constantly high oil prices. Therefore it argues that humanity does not have a moment to lose.

A Whole System Approach to design of our cities and transport systems will be vital to addressing this problem. Many sustainable transport experts argue that, to effectively reduce oil dependency in the transport sector, we need to transform our cities from their current automobile-dependant design to a more automobile-independent one. Therefore achieving sustainable transportation now, with what technology is

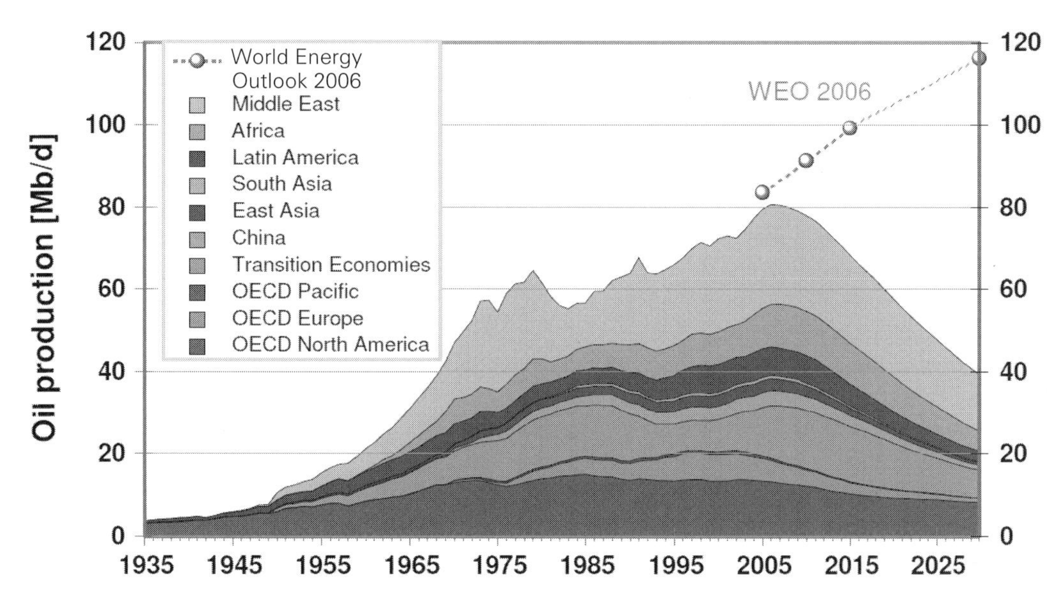

Source: Energy Watch Group (2007)[47]

Figure 1.3 *World oil production*

available, will require governments, business and citizens to work together to reduce their transportation needs through better urban/regional design and a shift to low-carbon-emitting transportation modes – especially through increased public transportation, rail, cycling and walking. Improvements in fuel efficiency of transportation vehicles (cars, trucks, buses and motorcycles) through Whole System Design approaches is also seen as a key strategy to reduce oil dependency.[48] Chapter 7, the Hypercar technical worked example, explains the benefits of a Whole System Approach to the design of cars. Many of the ideas outlined in the Hypercar are already being applied to hybrid cars, trucks, buses and motorbikes. Companies leading in this area are reaping significant financial benefits. Even as early as 2006, the Academy Awards car-park for Hollywood stars looked like a showroom for a hybrid car dealership. At that time, hybrids sold for as little as US$22,000 in the US. This is affordable for the average family, especially since such vehicles can cut the family fuel bill in half. In 2006, there was an eight-month wait for anyone wanting a hybrid in the US, such is their popularity.

As stated above, in October 2005, the world's largest retailer, Wal-Mart, announced a US$500 million climate change commitment, including initiatives to increase truck fleet fuel efficiency by 25 per cent in three years and double it in ten. And in England the first double-decker hybrid bus was launched in 2005 and already there are numerous hybrid motorbikes on the market.

Conclusion

Concern for these issues is not new. As far back as 1919, Svante Arrhenius, Director of the Nobel Institute, urged engineers to think of the next generation and embrace sustainable development:

> Engineers must design more efficient internal combustion engines capable of running on alternative fuels such as alcohol, and new research into battery power should be undertaken. ... Wind motors and solar engines hold great promise and would reduce the level of CO_2 emissions. Forests must be planted. ... To conserve coal, half a tonne of which is burned in transporting the other half tonne to market ... the building of power plants should be in close proximity to the mines. ... All lighting with petroleum products should be replaced with more efficient electric lamps. (Svante Arrhenius, 1926)[49]

Arrhenius called for the amount of waste from industry to be reduced so as to ensure that future generations could also meet their needs. He argued that the industrial world had given rise to a new kind of international warrior, who he called the 'Conquistador of Waste'. Arrhenius wrote eloquently:

> Like insane wastrels, we spend that which we received in legacy from our fathers. Our descendants surely will sensor us for having squandered their just birthright. ... Statesman can plead no excuse for letting development go on to the point where mankind will run the danger of the end of natural resources in a few hundred years.

A Whole System Approach to Sustainable Design will assist engineers to identify and design out waste in the first place and ensure that they play their part in achieving sustainable development. Hence Whole System Design offers exciting opportunities in which engineers can play their part to help companies, Australia and the world to achieve sustainable development in the 21st century.

Optional reading

Birkeland, J. (ed) (2002) *Design for Sustainability: A Sourcebook of Ecological Design Solutions*, Earthscan, London

Hargroves, K. and Smith, M. H. (2005) *The Natural Advantage of Nations: Business Opportunities, Innovation and Governance in the 21st Century*, Earthscan, London

Hawken, P., Lovins, A. and Lovins, L. (1999) *Natural Capitalism: Creating the Next Industrial Revolution*, Earthscan, London, www.natcap.org/sitepages/pid20.php, accessed 19 October 2007

Lovins, L. H. (2005) 'Green is good', *Sydney Morning Herald*, 19 April

McDonough, W. and Braungart, M. (2002) *Cradle to Cradle: Remaking the Way We Make Things*, North Point Press, New York

OECD (1998) *Eco-efficiency*, OECD, Paris

Porter, M. and van der Linde, C. (1995) 'Toward a new conception of the environment-competitiveness relationship', *Journal of Economic Perspectives*, vol IX–4, fall, pp97–118

Scheer, H. (2004) *The Solar Economy*, Earthscan, London

Van der Ryn, S. and Calthorpe, P. (1986) *Sustainable Communities: A New Design Synthesis for Cities, Suburbs and Towns*, Sierra Club Books, San Francisco, CA

Von Weizsäcker, E., Lovins, A. and Lovins, L. (1997) *Factor Four: Doubling Wealth, Halving Resource Use*, Earthscan, London

Notes

1. Hawken, P., Lovins, A. B. and Lovins, L. H. (1999) *Natural Capitalism: Creating the Next Industrial Revolution*, Earthscan, London.
2. An address to The International Society of Ecological Economists by the Federal Minister for the Environment and Heritage Senator the Hon Robert Hill, Australian National University, Canberra, 6 July 2000, www.deh.gov.au/minister/env/2000/sp6jul00.html, accessed 19 October 2007.
3. The Department of the Environment and Heritage (2001) *Product Innovation: The Green Advantage: An Introduction to Design for Environment for Australian Business*, Commonwealth of Australia, Canberra, www.environment.gov.au/settlements/industry/finance/publications/producer.html, accessed 7 May 2008.
4. The Department of the Environment and Heritage (2001) *Product Innovation: The Green Advantage: An Introduction to Design for Environment for Australian Business*, Commonwealth of Australia, Canberra, www.environment.gov.au/settlements/industry/finance/publications/producer.html, accessed 7 May 2008.
5. Porter, M. E. and van der Linde, C. (1995) 'Green and competitive: Ending the stalemate', *Harvard Business Review*, Boston, Reprint 95507.
6. Porter, M. and van der Linde, C. (1995) 'Toward a new conception of the environment-competitiveness relationship', *Journal of Economic Perspectives*, vol IX–4, fall, p126.
7. William McDonough and Partners (1992) *The Hannover Principles: Design for Sustainability*, William McDonough Architects, www.mcdonough.com/principles.pdf, accessed 19 October 2007.
8. See Rocky Mountain Institute – 'Natural capitalism' at www.rmi.org/sitepages/pid69.php, accessed 18 October 2007; Hawken, P., Lovins, A. and Lovins, L. (1999) *Natural Capitalism: Creating the Next Industrial Revolution*, Earthscan, London, www.natcap.org/sitepages/pid20.php, accessed 19 October 2007.
9. See Rocky Mountain Institute – 'Natural capitalism' at www.rmi.org/sitepages/pid69.php, accessed 18 October 2007; Hawken, P., Lovins, A. and Lovins, L. (1999) *Natural Capitalism: Creating the Next Industrial Revolution*, Earthscan, London, www.natcap.org/sitepages/pid20.php, accessed 19 October 2007.
10. See The Natural Step – 'What is sustainability?' at www.naturalstep.org/com/What_is_sustainability/, accessed 18 October 2007; Robert, K. H. (2002) *The Natural Step Story*, New Society, Gabriola Island, Canada.

11 US Environmental Protection Agency (n.d.) 'Twelve principles of green chemistry', www.epa.gov/green chemistry/pubs/principles.html, accessed 19 October 2007; Anastas, P. L. and Zimmerman, J. B. (2003) 'Design through the 12 principles of green engineering', *Environmental Science and Technology*, 1 March, ACS publishing, pp95–101.
12 Schmidt-Bleek, F. (1999) *Factor 10: Making Sustainability Accountable, Putting Resource Productivity into Practice*, p41, www.factor10-institute.org/pdf/F10REPORT.pdf, accessed 10 September 2007.
13 See Van der Ryn Architects – 'Five principles of ecological design' at http://64.143.175.55/va/index-methods. html, accessed 18 October 2007; van der Ryn, S. and Calthorpe, P. (1986) *Sustainable Communities: A New Design Synthesis for Cities, Suburbs and Towns*, Sierra Club Books, San Francisco, CA.
14 For detailed discussions on these criteria, readers are directed to Blanchard, B. S. (2004) *Logistics Engineering and Management* (sixth edition), Pearson Prentice-Hall, Upper Saddle River, NJ.
15 Hawken, P., Lovins, L. H. and Lovins, A. B. (1999) *Natural Capitalism: Creating the Next Industrial Revolution*, Earthscan, London, Chapter 6: 'Tunnelling through the cost barrier'.
16 Hawken, P., Lovins, L. H. and Lovins, A. B. (1999) *Natural Capitalism: Creating the Next Industrial Revolution*, Earthscan, London, Chapter 6: 'Tunnelling through the cost barrier'.
17 Hawken, P., Lovins, L. H. and Lovins, A. B. (1999) *Natural Capitalism: Creating the Next Industrial Revolution*, Earthscan, London, p121.
18 This programme has now become the Department of Resources, Energy and Tourism's Energy Efficiency Opportunities Program.
19 National Framework for Energy Efficiency and Department of Industry, Tourism and Resources (2006) *Energy Efficiency Opportunities: Assessment Handbook*, Commonwealth of Australia, Canberra, www.energy efficiencyopportunities.gov.au/assets/documents/energy efficiencyopps/EEO%20handbook%20screen2006110 2144033.pdf, accessed 7 March 2008; see also Department of Resources, Energy and Tourism, Energy Efficiency Best Practice (EEBP) programme, 'EEBP' in the 'X Sector Documents' at www.ret.gov.au/Programsandservices/EnergyEfficiencyBestPracticeEEB PProgram/Pages/default.aspx, accessed 7 March 2008.
20 Pears, A. (2004) 'Energy efficiency – Its potential: Some perspectives and experiences', background paper for International Energy Agency Energy Efficiency Workshop, Paris, April, www.naturaledgeproject.net/Documents/IEAENEFFICbackgroundpaperPearsFinal.pdf, accessed 30 March 2008.
21 Pears, A. (2004) 'Energy efficiency – Its potential: Some perspectives and experiences', background paper for International Energy Agency Energy Efficiency Workshop, Paris, April, www.naturaledgeproject.net/Documents/IEAENEFFICbackgroundpaperPearsFinal.pdf, accessed 30 March 2008.
22 The steam engine was invented in 1710 to pump water out of coal mines.
23 Christianson, G. (1999) *Greenhouse: The 200 Year History of Global Warming*, Walker & Company, New York.
24 Weizsäcker, E., Lovins, A. and Lovins, H. (1997) *Factor Four: Doubling Wealth, Halving Resource Use*, Earthscan, London; McDonough, M. and Braungart, M. (2002) *Cradle to Cradle – Remaking The Way We Make Things*, North Point Press, New York, www.mcdonough.com/cradle_to_cradle.htm; Birkeland, J. (2002) *Design for Sustainability*, Earthscan, London.
25 Birkeland, J. (2005) *Design for Ecosystem Services – A New Paradigm for Eco-design*, International Sustainable Buildings Conference, Tokyo; Birkeland, J. (ed) (2002) *Design for Sustainability: A Sourcebook of Ecological Design Solutions*, Earthscan, London.
26 McDonough, M. and Braungart, M. (2002) *Cradle to Cradle – Remaking The Way We Make Things*, North Point Press, New York.
27 Porter, M. E. and van der Linde, C. (1995) 'Green and competitive: Ending the stalemate', *Harvard Business Review*, Boston, Reprint 95507, p122.
28 Pears, A. (2004) 'Energy efficiency – Its potential: Some perspectives and experiences', background paper for International Energy Agency Energy Efficiency Workshop, Paris, April 2004, www.naturaledgeproject.net/Documents/IEAENEFFICbackgroundpaperPears Final.pdf, accessed 30 March 2008.
29 Hargroves, K. J. and Smith, M. H. (2005) *The Natural Advantage of Nations: Business Opportunities, Innovation and Governance in the 21st Century*, Earthscan, London, p154.
30 Pears, A. (2004) 'Energy efficiency – Its potential: Some perspectives and experiences', background paper for International Energy Agency Energy Efficiency Workshop, Paris, April 2004, www.naturaledgeproject.net/Documents/IEAENEFFICbackgroundpaperPears Final.pdf, accessed 30 March 2008.
31 Fukasaku, Y. (1999) 'Stimulating environmental innovation', *The STI Review*, vol 25, no 2, Special Issue on Sustainable Development, OECD, Paris.
32 According to the OECD, the term 'decoupling':

has often been used to refer to breaking the link between the growth in environmental pressure associated with creating economic goods and services. In particular it refers to the relative growth rates of a pressure on the environment and of the economically relevant variable to which it is causally linked. Decoupling occurs when growth rate of the environmentally relevant variable is less than that of its economic variable (e.g. GDP) over a period of time.

33 See UN Millennium Ecosystem Assessment at www.maweb.org/en/index.aspx, accessed 28 March 2008
34 OECD (1998) *Eco-efficiency*, OECD, Paris, p71.
35 Netherlands Environmental Assessment Agency (2007) *Environmental Balance 2007*, Netherlands Environmental Assessment Agency (MNP), Bilthoven, The Netherlands.
36 See *Organisation for Economic Co-operation and Development* – 'Draft Agenda 2001' at www1.oecd.org/env/min/2001/agenda.htm, accessed 18 March 2008.
37 VandeHei, J. (2005) 'President holds firm as G-8 summit opens: Bush pledges to help Africa, but gives no ground on environmental policy', *Washington Post*, 7 July 2005, pA14, www.washingtonpost.com/wp-dyn/content/article/2005/07/06/AR2005070602298.html, accessed 20 August 2008.
38 Lovins, L. H. (2005) 'Green is good', *Sydney Morning Herald*, 19 April.
39 DEH: Greenhouse Office (2005) *Tracking to the Kyoto Target 2005*, DEH, www.greenhouse.gov.au/projections/pubs/tracking2005.pdf, accessed 18 March 2007.
40 IPCC (2001) *Climate Change 2001: Synthesis of the Third Assessment Report*, Intergovernmental Panel on Climate Change, United Nations Environment Program/World Meteorological Organisation, Cambridge University Press.
41 Peatling, S. (2004) 'Carbon emissions must be halved, says science chief', *Sydney Morning Herald*, 19 July.
42 Office of the Press Secretary (2006) 'President Bush delivers State of the Union Address', Office of the Press Secretary, www.whitehouse.gov/news/releases/2006/01/20060131-10.html, accessed 18 March 2008.
43 Index Mundi website: *Crude Oil (petroleum) Monthly Price* at http://indexmundi.com/commodities/?commodity=crude-oil&months=300, accessed 5 September 2008.
44 In late May 2005, Robert Hirsch presented the substance of the report at the annual workshop of the Association for the Study of Peak Oil (ASPO) in Lisbon, Portugal, to an audience of about 300, www.cge.uevora.pt/aspo2005/abscom/Abstract_Lisbon_Hirsch.pdf, accessed 18 March 2008.
45 Johnson, H. R., Crawford, P. M. and Bunger, J. W. (2004) *Strategic Significance of America's Oil Shale Resource: Volume I – Assessment of Strategic Issues*, US Department of Energy, Washington, DC, p10, www.fossil.energy. gov/programs/reserves/npr/publications/npr_strategic_significancev1.pdf, accessed 18 March 2008.
46 Hirsch, R. L., Bezdek, R. and Wendling, R. (2005) *Peaking of World Oil Production: Impacts, Mitigation and Risk Management*, US Department of Energy, National Energy Technology Laboratory, www.hilltoplancers.org/stories/hirsch0502.pdf, accessed 19 October 2007.
47 Energy Watch Group (2007) *Crude Oil: The Supply Outlook, Report to the Energy Watch Group*, Energy Watch Group, p68, www.energywatchgroup.org/fileadmin/global/pdf/EWG_Oilreport_10-2007.pdf, accessed 23 July 2008.
48 Lovins, A. B., Datta, E. K., Bustnes, O. E., Koomey, J. G. and Glasgow, N. J. (2004) *Winning the Oil Endgame: Innovation for Profits, Jobs and Security*, Book and Technical Annexes, Rocky Mountain Institute, Snowmass, CO, www.oilendgame.com, accessed 29 July 2007.
49 Arrhenius, S. (1926) *Chemistry in Modern Life*, Van Nostrand Company, New York, NY.

2

The Fundamentals of Systems Engineering to Inform a Whole System Approach

Educational aim

Chapter 2 provides an introduction to conventional Systems Engineering so that we can show in Chapters 3–5 how a Whole System Approach to Sustainable Design will enhance the discipline. In Chapter 1, we introduced the fact that many engineered systems are sub-optimally designed, because engineers have not taken the time to optimize the whole system. Chapter 1 described how this fact has inspired the field of Whole System Design. Chapter 2 now shows that this fact has also inspired Systems Engineering. It first highlights the similarities between some of the principles and motivations of good Systems Engineering and Whole System Design, before outlining the differences: a Whole System Approach to Sustainable Design covers more than simply engineering design. Whole System Approaches to Sustainable Design can be applied to all fields of design – by architects and industrial, urban and landscape designers, not just by engineers. Whole System Design is also different from traditional Systems Engineering in that it has been more focused on better whole system optimization to go beyond simply better efficiencies and achieve ecological sustainability. This key difference is highlighted in Chapters 3–5, with Chapter 3 illustrating how Whole System Design enhances traditional Systems Engineering with its greater emphasis on ecological sustainability in the design process. Chapter 2 also overviews key terminology and concepts derived from the field of systems science that are relevant to systems engineers and whole system designers. It is important to put Whole System Design Approaches in the context of traditional Systems Engineering to assist the rapid mainstreaming of the latest insights from Whole System Design into engineering design courses and practices. Also, traditional Systems Engineering will be greatly enriched by integrating it with the latest insights from the Whole System Design literature.

Required reading

Blanchard, B. S. and Fabrycky, W. J. (2006) *Systems Engineering and Analysis* (fourth edition), Pearson Prentice Hall, Upper Saddle River, NJ, Chapter 1, pp2–21

Honour, E. C. (2004) *Understanding the Value of Systems Engineering*, proceedings of the Fourteenth Annual Symposium of the International Council on Systems Engineering, Toulouse, France, pp1–16, www.incose.org/secoe/0103/ValueSE-INCOSE04.pdf, accessed 5 October 2007

Rocky Mountain Institute (1997) 'Tunnelling through the cost barrier', *RMI Newsletter*, summer, pp1–4, www.rmi.org/images/other/Newsletter/NLRMIsum97.pdf, accessed 5 October 2007

Introduction: Whole system design and systems engineering

In Chapter 1, the benefits of Whole System Design (WSD) were outlined. Chapter 1 and the books and reports referenced within it show that, historically, many engineered systems have not gone through a rigorous WSD optimization process. For example, as the Rocky Mountain Institute[1] write, most energy-using technologies are designed in ways that are intended to produce an optimized design, but actually produce sub-optimal solutions:

1 Components are optimized in isolation from other components (potentially 'pessimizing' the systems of which they are a part).
2 Optimization typically considers single rather than multiple benefits.
3 The optimal sequence of design steps is not usually considered.

Chapter 1 showed that pursuing a Whole System Approach to Sustainable Design can help engineers to achieve significant resource efficiency and productivity gains and thus help reduce pressures on the environment. An effective WSD optimization, carried out in the early stages of design projects, provides significant economic, social and environmental benefits. Decisions made early in the design process have an enormous impact on life-cycle system costs, both economic and environmental.[2] Figure 2.1 shows that approximately 60 per cent of life-cycle costs are determined in the concept phase (Need Definition and Conceptual Design), and a further 20 per cent are determined in the design phase (Preliminary Design and Detail Design).

In addition to the direct costs associated with the project, the cost of making design changes escalates as

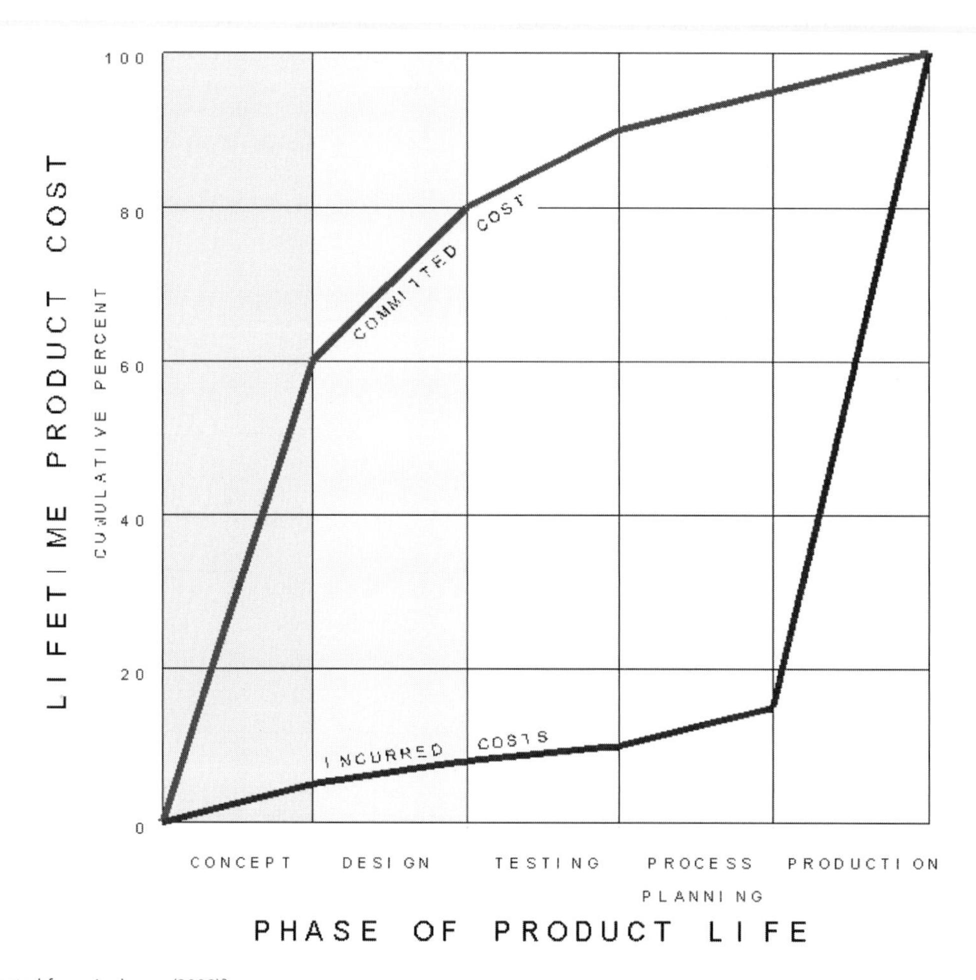

Source: Adapted from Andersen (2008)[3]

Figure 2.1 *Comparison of the incurred costs and committed costs for each phase of system development*

system development progresses. Figure 2.2 shows that the cost of making design changes is lowest during the initial design phase, is 10 times higher during the pre-production phase and more than 80 times higher during the production phase.

These facts have led many in the design professions to call for greater effort to be made in the concept and early design phases – known as *Front End Loading*. There is tremendous leverage in investing adequate human and financial resources into the earliest phases of the development process. A Front End Loading can lead to better considered decisions, lower life-cycle costs and fewer late changes, through a concentration of design activity and decisions in the earliest phases, where changes cost the least. This emphasis on more Front End Loading makes intuitive sense, as shown in Figure 2.3. In traditional design, without consideration of whole system approaches and the life-cycle of a product or development, the creation of a system is focused on production, integration and testing. In a Whole System Approach to Sustainable Design process, greater emphasis on Front End Loading creates easier, more rapid integration and testing by avoiding many of the problems normally encountered in these phases. By reducing risk early in the design process, the overall result is a saving in both time and cost, with a higher quality system design. There are now a range of empirical studies that support the idea that increasing the level of Systems Engineering has a positive effect on cost compliance and the quality of the project.[4]

One of the reasons why so many technical systems are not based on a Whole System Approach to Sustainable Design is because the engineering and design professions have become highly specialized. As the field of engineering has grown exponentially, there has been a need for engineers to specialize, as no one person can now master all the individual fields of engineering. Through that process of specialization over the last

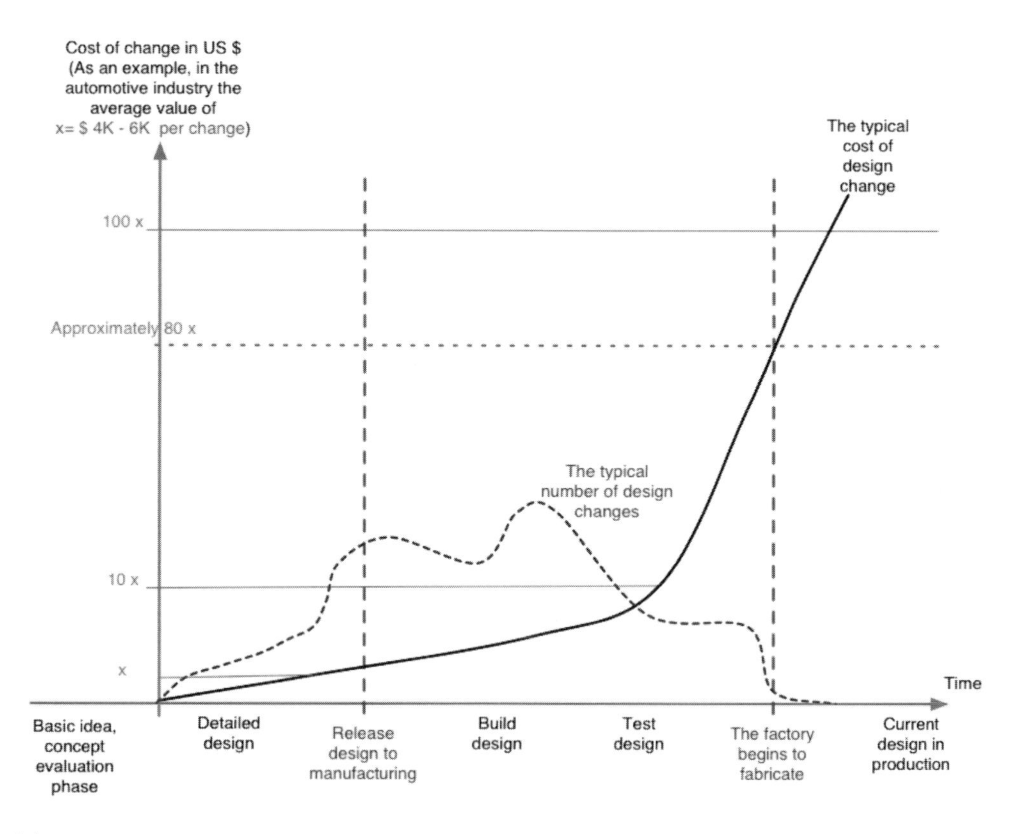

Source: Ranky[5]

Figure 2.2 *The cost of making design changes throughout each phase of system development*

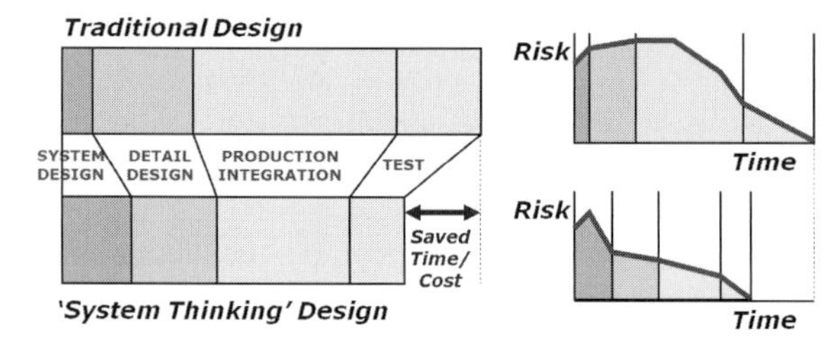

Source: Honour (2004)[6]

Figure 2.3 *The value of Front End Loading in reducing costs and risks*

century, many engineers have lost the Victorian[7] engineering art of a multi-disciplinary whole system optimization, simply because they do not know enough technical detail to individually complete a true whole system optimization. While engineers have become more and more specialized, modern technologies have become increasingly complex. As Blanchard and Fabrycky comment in their textbook on Systems Engineering:[8]

> Although engineering activities in the past have adequately covered the design of various system components (representing a bottom–up approach), the necessary overview and understanding of how these components effectively perform together is frequently overlooked.

Whole system designers like Amory Lovins, Hunter Lovins, Ernst von Weizsäcker, Bill McDonough, John Todd, Janis Birkeland and Alan Pears (see Optional reading, p40) have recognized the desperate need for designers to be able to step back and analyse the whole system to ensure that the solution is as effective as possible. It is also important for engineers and designers not only to think of their own scope of work in a whole system manner, but also to understand how their expertise can be optimized within the context of the WSD team as well. Chapter 1 described how the field of WSD was developed to address these issues. As well as WSD, a new field of engineering has been created, called 'Systems Engineering' to also address these issues.

The field of Systems Engineering, like WSD, has arisen out of the recognition of:

- The need for better Front End Loading; and
- The need for engineering designs to optimize the whole system using a life-cycle approach.

The field of Systems Engineering, like WSD, has been created out of the recognition that any changes to the design of sub-systems affect the overall system design and performance. Systems Engineering has been created by the engineering profession out of recognition that, as engineering has grown more sophisticated and complex, it has become necessary to focus more on managing carefully how the engineering of components affects the overall system design. Done well, Systems Engineering ensures that the whole is greater than the sum of the parts, just as WSD does. Systems Engineering is the traditional field of engineering which helps engineers understand how to optimize an entire system.

However, rarely do Systems Engineering textbooks emphasize ecological sustainability as a key goal to be included in the daily practise of Systems Engineering. A Whole System Approach to Sustainable Design is also different from traditional Systems Engineering in that it is more focused on whole system optimization not only for efficiencies, but also for ecological sustainability. These key elements of a Whole System Approach to Sustainable Design are highlighted in Chapters 3–5 to demonstrate how it can enhance traditional Systems Engineering. Also, a Whole System Approach to Sustainable Design is a broader concept than Systems Engineering, since the Whole System Approach field is relevant for many professionals – architects and industrial, urban and landscape designers, not just engineers. We believe it is vital that engineers understand this difference in order to appreciate how the latest insights from the Whole System Approach field (see optional reading) complement and enhance traditional Systems Engineering to help them focus on achieving ecologically sustainable outcomes.

Chapter 2 and the start of Chapter 3 provide an overview of traditional Systems Engineering. This is done so that the second half of Chapter 3 and Chapters 4 and 5 can highlight how the latest operational insights from a Whole System Approach to Sustainable Design can enhance the operational implementation of Systems Engineering principles to achieve more sustainable outcomes. We believe that it is also important to put a Whole System Approach to Sustainable Design in the context of traditional Systems Engineering to assist the rapid mainstreaming of the latest insights from leading thinkers in the field of WSD for sustainability, such as Amory Lovins, Ernst von Weizsäcker, Bill McDonough, Janis Birkeland and Alan Pears (see optional reading). This is a key goal of Chapters 2–5.

Also, we believe Systems Engineering is greatly enriched by integrating it with the large body of work on a Whole System Approach to Sustainable Design (see optional reading). There are many Systems Engineering success stories, like the hybrid car, which can help society achieve ecological sustainability, and yet these are not covered in most Systems Engineering textbooks, which seem to overlook case studies that apply the Systems Engineering methodology to improving the environmental performance of designs. Chapters 6–10 of this volume provide engineering practitioners, lecturers and students with detailed technical worked examples of WSD for sustainability that could be both included in Systems Engineering textbooks and taught in Systems Engineering and Systems Design courses around the world.

What is systems engineering?

Systems Engineering is a process whereby engineers analyse and optimize the whole technical system, which is composed of components, attributes and relationships, to achieve a specified goal. Components, attributes and relationships, in an engineering sense, are defined as follows:

- *Components* are the operating parts of a system (see Figure 2.4), consisting of input, process and output. Each system component may assume a variety of values to describe a system state set by control actions and restrictions.
- *Attributes* are the properties or discernible manifestations of the components of a system. These attributes characterize the system.
- *Relationships* are the links between components and attributes.

As Blanchard and Fabrycky explain:[9]

> A system is a set of inter-relating components that form an integrated whole with a common goal or purpose. In engineering, the objective or purpose of a system must be explicitly defined and understood so that system components may be selected to provide the desired outcome. The purposeful action performed by a system is referred to as its function. Common system functions include those of transforming and altering material, energy and/or information. Systems that alter material, energy or information are composed of structural components, operating components and flow components. Structural components are the static parts, operating components are

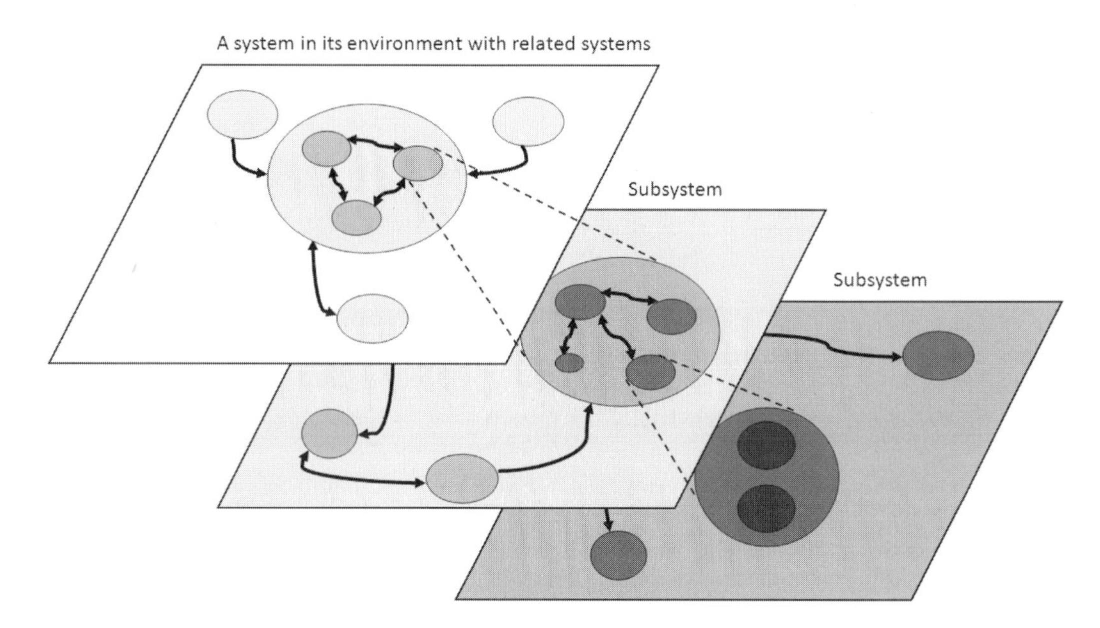

Source: Adcock (n.d.)[10]

Figure 2.4 *The composition of a system*

> the parts that perform the processing, and flow components are the material, energy or information being altered. Every system is made up of components and any component can be broken down into smaller components.

Systems engineers usually work with engineers from all the traditional engineering disciplines to optimize the whole system to achieve a defined goal or purpose. Systems Engineering plays the role of integrating all the fields of engineering to achieve still greater results (see Figure 2.5). Blanchard and Fabrycky sum this up as follows:[11]

> Systems Engineering involves an interdisciplinary or team approach throughout the system design and development process to ensure that all design objectives are addressed in an effective and efficient manner. This requires a complete understanding of many different design disciplines and their inter-relationships, together with the methods, techniques and tools that can be applied to facilitate implementation of the system engineering process.

In Systems Engineering, as with any engineering discipline, the objective or purpose of the system must be explicitly defined and understood to ensure that an effective solution is designed. Once the *purpose* is defined, this allows the engineer to determine the best way to meet a desired outcome. There are almost always several different ways to engineer a solution to meet a specified need or service. It is up to the systems engineer to conceive of and work on alternative ways to meet these needs and provide these services. It is the role of the good engineer or designer to determine which of these alternatives is the optimal way to provide a service and meet society's needs. As Blanchard and Fabrycky write:[12]

> A better and more complete effort is required regarding the initial definition of system requirements, relating these requirements to specific design criteria and the follow-on analysis effort to ensure the effectiveness of early decision-making in the design process. The true system requirements need to be well defined and specified, and the traceability of

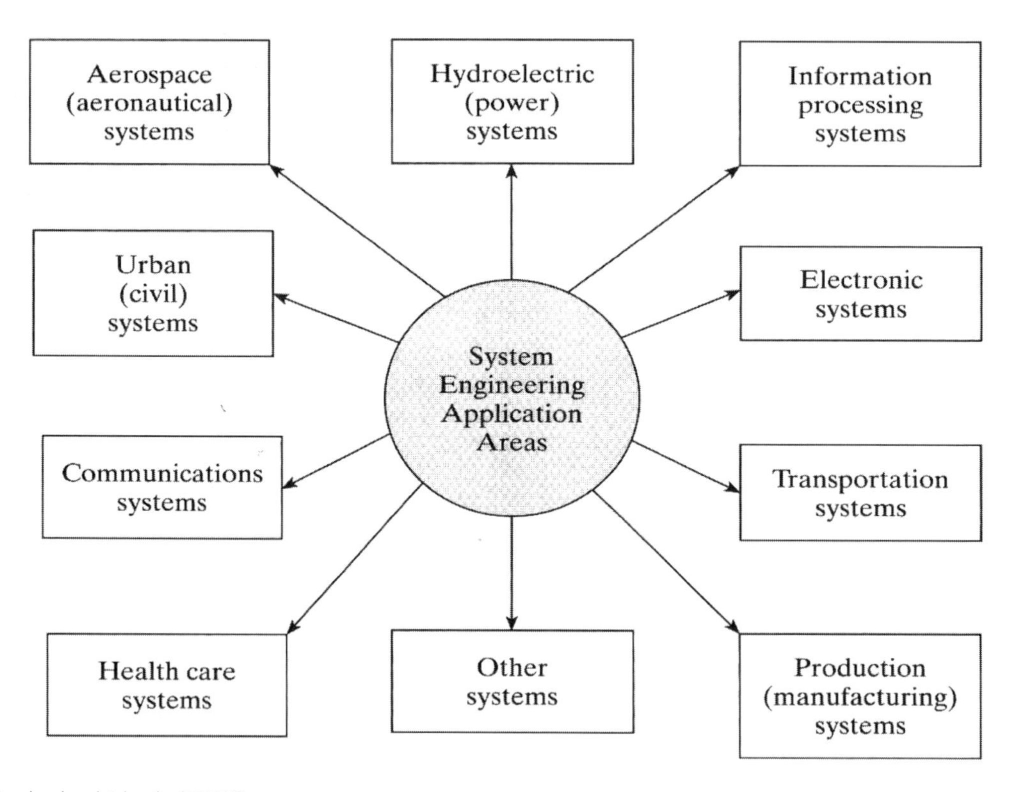

Source: Blanchard and Fabrycky (2006)[13]

Figure 2.5 *Application areas for System Engineering*

these requirements, from the system level down, needs to be visible. In the past, the early 'front-end' analysis was minimal. The lack of defining an early 'baseline' has resulted in greater individual design efforts downstream.

Taking a Systems Engineering approach helps ensure that engineers examine the many choices that are available to meet the specific needs of society, with each approach having its own unique energy and material needs and environmental impacts. Energy and materials are not used for their own sake. They are inputs into a system that provides a function that is considered useful or valuable by society. The client or customer wants cold beer and warm showers, not kilowatts of energy. People want to drink out of something hygienically packed and easy to handle, and don't so much want to use a container that creates a waste problem. People want mobility, they wish to get from A to B, but don't necessarily want more congestion from cars. They want the services that energy, materials and information provide, not the environmental costs and by-products that they can inadvertently create. This means that there are numerous ways that engineers can provide these services whilst dramatically reducing the environmental impacts of the energy and materials used to provide them. Taking a services perspective can free engineers to create totally new ways of meeting people's everyday needs.

Systems Engineering emphasizes the importance of stepping back from the problem and asking crucial questions to ensure that the most appropriate solutions are found. Hitchins's list of Systems Engineering tenets[14] serves as a general guide for effective Systems Engineering:

1 Approach an engineering problem with the highest level of abstraction for as long as practicable.
2 Apply 'disciplined anarchy' – that is, explore all options and question all assumptions.
3 Analyse the whole problem breadth-wise before exploring parts of the solution in detail; understand the primary system level before exploring the sub-system.
4 Understand the functionality of the whole system before developing a physical prototype.

Chapter 3 will show how Systems Engineering can be enhanced to incorporate sustainability considerations and hence encourage the development of sustainable systems.

In the past, due to engineers not considering a wide range of options, some engineering applications have performed poorly as part of the larger system. This is partly due to a lack of knowledge beyond one's own engineering discipline and a lack of knowledge amongst designers of natural systems and their limits and thresholds. Confidence in the intrinsic value of technological progress has also led at times to scientific and engineering designers being too quick to reach their conclusions. There has been an under-appreciation of the value of a precautionary approach to technological development. Two examples that illustrate this were the development of leaded petrol and ozone destroying CFCs for air-conditioning and refrigerators.

Thomas Midgley, the chief engineer responsible for the decision to add lead to petrol[15] and to use chlorofluorocarbons (CFCs)[16] for numerous industrial and consumer applications, did not appreciate the ecological effects of heavy metals and certain chemicals. Midgley died believing that CFCs were of great benefit to the world, and a great invention. He was not the only expert to be guilty of ignorance. Almost all scientists and engineers until the 1950s were ignorant of the negative environmental effect of burning fossil fuels. All assumed that the oceans and forests would absorb all the carbon dioxide produced from burning fossil fuels, and it never occurred to them that this human behaviour could be a problem. The reason plastics do not degrade in the environment is because they are designed to be persistent; similarly fertilizers were designed to add nitrogen to soil, so it is not an accident that they also add nitrogen to waterways, thus leading to algae blooms. Part of the problem, as argued by Commoner in his book *The Closing Circle*,[17] is that designers make their aims too narrow. Commoner argued that historically designers have seldom aimed to protect the environment, but that technology can be a successful part of the Earth's natural systems, 'if its aims are directed towards the system as a whole rather than some apparently accessible part'. Commoner advocated a new type of technology that is designed with the full knowledge of ecology and the desire to fit in with natural systems.

A lack of appreciation of the need to take the broader environmental and social systems approach when addressing problems has been an issue not only in engineering, but also in many other disciplines, such as medicine. The following case study illustrates well what can go wrong when the broader system is not taken into consideration when designing solutions to problems, effectively treating the symptoms but not creating lasting solutions.

Why an understanding of systems matters

Case study: Operation Cat Drop

In the 1950s in Borneo, malaria was identified as a significant health issue. In response to this problem, the World Health Organization (WHO) decided to take measures to significantly reduce the mosquito population, since mosquitoes are carriers of malaria. To achieve this they used the insecticide DDT, which effectively reduced mosquito populations and significantly reduced the incidence of malaria. However, the WHO failed to appreciate the full scope of their actions. DDT not only successfully killed mosquitoes, it also attacked a parasitic wasp population. These wasps had kept in check the population of thatch-eating caterpillars. So with the unforeseen removal of the wasps, the caterpillar population blossomed, and soon thatch roofs started falling all over Borneo.

There were additional unforeseen effects. Insects poisoned by DDT were consumed by geckoes. The biological half-life of DDT is around eight years, so animals like geckoes do not metabolize it very fast, and it stays in their system for a long time. The geckoes carrying the DDT poison were in turn hunted and eaten by the cat population. With more cats dying prematurely, rats took over and multiplied, and this in turn led to outbreaks of typhus and sylvatic plague (which are passed on by rats). At this stage the effects of the intervention on the health of the people of Borneo were worse than the original malaria outbreak. So the World Health Organization (WHO) resorted to the extraordinary step of parachuting cats into the country. The event has become infamously coined 'Operation Cat Drop'.[18]

The WHO had failed to consider the full implications of their actions on the delicate natural systems of Borneo. Because they lacked understanding of the basic effects of DDT (now banned in many countries), a high cost was paid for this mistake. By considering only the first-level relationship between mosquitoes as carriers of malaria and humans as recipients of malaria, the WHO unrealistically assumed that this relationship could be acted upon independently of any other variables or relationships. They considered one aspect of the system, rather than the whole system (the entire ecology).

This example demonstrates the importance of a Whole System Approach to challenges/problems in seeking sustainable (lasting) solutions. In the real world, one relationship strand (for example, mosquito–human) cannot be separated from the rest of the system. All of the parts of the system are tied together in a complex fabric, and changing one part of the system can lead to profound changes throughout the rest of the system which may not at first glance appear at all connected to the point of action.

Broadening the problem definition

Systems Engineering has evolved out of this understanding of the need to consider the complex inter-relationships of systems. Changes which are seemingly narrow in scope can set off a domino effect that reaches much wider than ever anticipated.

Systems Engineering recognizes that systems exist throughout the natural and man-made world, wherever there is complex behaviour arising from the interaction between things. This behaviour can only be understood by considering 'complete systems' as they interact within their 'natural' environment. The goal of Systems Engineering is to consider the whole system, in its environment, through its whole life-cycle (see Figure 2.6). The viability of an engineered system, design or product generally relies upon interactions outside of its immediate boundary. Systems Engineering simultaneously focuses on the specific product to be designed while considering how that product fits within the context of one or more 'containing systems', including the natural environment.

To solve complex 'System Problems', we must engineer complete 'System Solutions' through a combination of:

- The ability to understand, describe, predict, specify and measure the ways in which elements of an engineered system will affect elements of a complex system;
- The ability to apply 'traditional' engineering knowledge to create, modify or use system elements to manipulate, maintain or enhance the resilience of the complex system; and

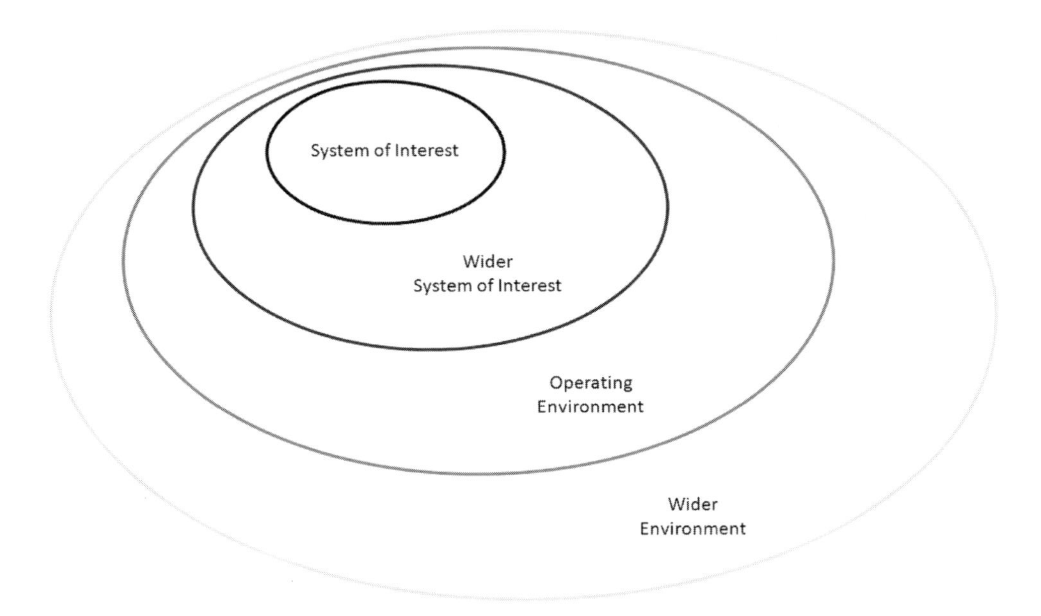

Source: Adcock (n.d.)[19]

Figure 2.6 *A system and the many layers of its environment*

- The ability to organize, manage and resource projects in such a way as to achieve the above aims, within realistic constraints of cost, time and risk.

It is vital, with the Earth's ecosystems having already lost so much of their resilience and now under increasing environmental pressures, that engineers in the 21st century ensure their engineering solutions do not create new, unforeseen problems which further add to environmental pressures. Before discussing in Chapter 3 the detailed operational steps of conventional Systems Engineering and how this can be enhanced through a Whole System Approach to Sustainable Design, it is important to overview some systems definitions and concepts to provide a foundation for the rest of the book. The rest of this chapter therefore introduces some of the key terminology of systems analysis and Systems Engineering. Chapters 3–5 then discusses the key operational process steps of good Systems Engineering and how these can be enhanced by the 10 Elements of a Whole System Approach to Sustainable Design. Chapters 6–10 then provide more detailed technical worked examples to demonstrate further the value of a WSD approach.

What is a system?

Systems are everywhere. Our universe, the Earth, even a tiny atom is a system. But only very recently has humanity started to engineer human-made systems. And only in the last few hundred years has humanity begun to truly understand the detailed workings, laws and relationships of both natural and human-made systems. We have all heard of various forms of technological systems: computer systems, security systems and manufacturing systems, for example. But what do we actually mean when we describe something as a 'system'?

> A system is an open set of complementary, interacting parts, with properties, capabilities and behaviours emerging both from the parts and from their interactions. Hence changing one part of the system will ultimately have an effect on the performance of other parts in the system.

Blanchard and Fabrycky define a system as follows:[20]

> A system is any combination of elements or parts forming a complex of unitary whole, such as a river system or a transportation system; any assemblage or set of correlated members, such as a system of currency; an ordered and comprehensive assemblage of facts, principles or doctrines in a particular field of knowledge or thought, such as a system of philosophy; a coordinated body of methods or a complex scheme or plan of procedure, such as a system of organization and management; or any regular or special method or plan of procedure, such as a system of marking, numbering or measuring. Not every set of items, facts, methods or procedures is a system. A random group of items ... would constitute a set with definite relationships between the items, but it would not qualify as a system, because there is an absence of unity, functional relationship and useful purpose.

In analysing and developing systems, it is important to establish system boundaries. Strategies to establish system boundaries vary. Typically, the wider the boundaries, the greater the opportunities to influence system performance, service delivery, environmental impact and cost-effectiveness. In this volume, where the focus is primarily on analysing and developing technical engineered systems for environmental sustainability, the boundaries generally encompass:

- All subsystems involved in developing, operating, maintaining and retiring the system that can be directly influenced;
- All other subsystems involved delivering the system's services that can be directly influenced; and
- The interactions between the subsystems within the boundaries.

In addition, analysing and developing systems actively considers the subsystems and interactions beyond the boundaries, particularly those related to the creation and delivery of input resources at the system boundaries and those related to the processing of output resources at the system boundaries.

As an example of establishing a system boundary, consider the (simplified) task of an automotive original equipment manufacturer (OEM) developing a modern internal combustion engine for a car. The development process typically involves selecting and integrating engine components produced by external suppliers, and assembling the engine. In relation to Figure 2.6, the relevant systems are:

- System of interest: engine; assembly process for the engine; procedures for maintenance and end-of-life processing; spare components;
- Wider system of interest: production processes for engine components and raw materials; infrastructure for fuel access, maintenance, spare components access and end-of-life processing; car; assembly process for the car;
- Operating environment: car; local climate; and
- Wider environment: roads; urban and built environment; biosphere.

If the task is strictly limited to engine development, then the system boundaries encompass only the subsystems listed in 'system of interest'. These subsystems' components will be selected, modified and manipulated to optimize the system. For example, the most suitable crankshaft will be selected by the OEM's designer from a pool of crankshafts manufactured by various suppliers – thus, the engine development process has a direct influence on the performance of the crankshaft. Engine development also actively considers the 'wider system of interest', 'operating environment' and 'wider environment'. These subsystems will not be modified by the engine development process, but the subsystems may be modified in response to demand for their services. For example, the production process used by suppliers for engine crankshafts will not be modified by the designer, but may be modified by the suppliers to consume less energy if the OEM has committed to reducing its greenhouse gas emissions – thus, the engine development process has an indirect influence on the production process for crankshafts.

Expanding the boundaries to encompass the whole car will increase the opportunities to develop a better car (including a better engine) by granting the designer access to directly select, modify and manipulate a wider variety of subsystems and components. Chapter 7 presents a worked example of designing a car with the system boundaries at the level of the car. Chapters 6, 8, 9, and 10 present similar worked examples for other technical engineered systems.

Systems analysis

Increasingly, engineers are being asked to analyse and address complex systems problems, such as traffic congestion, climate change and urban water management.[21] Many of the sustainability challenges faced by society involve complex interactions between the technical, social and economic dimensions. An ability to undertake systems analysis can help engineers tackle the complexity of real world problems with greater confidence, and there is an extensive field of such systems analysis which engineers can turn to for ideas on how to tackle complex systems problems.[22]

Analysis of systems involves an investigation of the multiple relationships of elements that comprise a system. Systems analysis uses diagrams, graphs and pictures to describe and structure inter-relationships of elements and behaviours of systems. Every element in a system is called a *variable*, and the influence of one element on another element is called a *link*; this can be represented by drawing an arrow from the causing element to the affected element. In analysis of systems, links always comprise a 'circle of causality' or a *feedback loop*, in which every element is both cause and effect. For example, take the urban expansion/induced traffic issue depicted below (Figure 2.7). To relieve traffic congestion in cities (variable #1), freeways are added or extended (variable #2). By adding more/extending freeways, people are able to live further out from the city, and hence more residential properties are built further out from the city (variable #3). More people living further out means more people drive into the city via the new freeways, hence contributing even more to the traffic congestion problem (feedback loop).

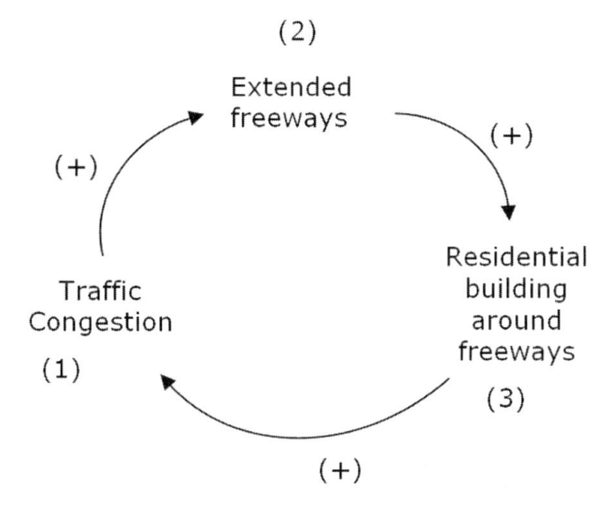

Figure 2.7 *Variables, links and feedback loops applied to the issues of urban expansion and induced traffic*

However, it should be recognized that the variables in a systems diagram, such as Figure 2.7, don't occur in series. In reality, all of these events occur simultaneously, which further places emphasis on the interconnected relationship between variables. There are two ways to represent feedback systems – as *reinforcing loops* or *balancing loops*.

Reinforcing loops

Reinforcing loops generate exponential growth and then collapse. As just described, an example of reinforcing loops is urban expansion/induced traffic in many western cities. Several studies confirm just how quickly urban expansion/induced traffic can take over landmass, as the Sierra Club explain:

> Shortly after the lanes or road is opened traffic will increase to 10 to 50% of the new roadway capacity as public transit or carpool riders switch to driving, or motorists decide to take more or longer trips or switch routes. This is short-term induced travel. In the longer term (three years or more), as the new roadway capacity stimulates more sprawl and motorists move farther from work and shopping, the total induced travel rises to 50 to 100% of the roadway's new capacity. This extra traffic clogs local streets at both ends of the highway travel.[23]

The expansion of several US cities is visible on satellite images by the US Geological Survey.[24] This expansion occurs despite some of these cities having politically-defined urban growth boundaries in place to control urban expansion. It is important to note that, in contrast to western cities, rapid road development in Chinese cities was in anticipation of increased traffic as the country became more 'modernized' and industrialized, and that urban development did not necessarily follow road development. Such urban expansion, as in Chengdu, China, is also visible on satellite images by NASA.[25]

Reinforcing loops, by definition, are incomplete. Somewhere, sometime, it will encounter at least one balancing mechanism that limits the spiralling up or spiralling down effect.

Balancing loops

Balancing loops are forces of resistance that balance reinforcing loops. They can be found in nature (chemical buffers in oceans or cellular organizations) and indeed other systems, and are the processes that fix problems and maintain stability. An example of balancing loops in engineering systems is the suspension system in an automobile. The suspension system is designed to cushion and control disturbances to the height of the passenger cabin. While a disturbance will initially change the cabin's height, the suspension system will eventually restore the cabin to its original height. Systems that are self-regulating or self-correcting comprise of balancing loops. Balancing processes are bound to a constraint or target which is often set by the forces of the system, and will continue to add pressure until that target has been met.

A significant characteristic of many systems, and often the most ignored, is delay. Delays in loops occur when a link takes a relatively long time to act out, and can have an enormous influence on a system, often exaggerating the behaviour of parts of the system and hence the general behaviour of the whole system. Delays are subtle and often neglected, yet they are prevalent in systems and must be actively considered. Delayed effects are very common in natural systems. This is fundamentally one of the reasons why we currently have the loss of resilience globally of many of the Earth's ecosystems, as highlighted by the UN Millennium Ecosystem Assessment.[26] Delayed effects of humanity's pressure on the environment mean that we can overshoot ecological system thresholds without knowing it. This has been a major factor in lulling humanity in general and designers in particular into a false sense of security that things are 'not that bad' environmentally.

Natural systems often exhibit delayed feedbacks: The problem of overshoot

Over the last two centuries, scientists have researched and begun to understand complex natural systems. They have found that the inherent resilience of natural systems means that they often exhibit a delayed feedback to environmental pressures. It is therefore often difficult to simply see with the naked eye how pollution and development are reducing the resilience of natural ecosystems *until it is too late* and the ecological system has been pushed past a particular irreversible threshold. Jared Diamond showed in his 2005 book *Collapse*[27] that this delayed feedback was a factor in the collapse of many past civilizations. Richard St. Barbe Baker, renowned UK forester and founder of Men of Trees in the 1920s, was one of the first to draw the modern world's attention to the risks that arise from the fact that natural systems often exhibit delayed feedback:

> The great Empires of Assyria, Babylon, Carthage and Persia were destroyed by floods and deserts let loose in the wake of forest destruction. Erosion following forest destruction and soil depletion has been one of the most powerfully destructive forces in bringing about the downfall of civilizations and wiping out human existence from large tracts of the Earth's surface. Erosion does not march with a blast of trumpets or the beating of drums, but its tactics are more subtle, more sinister. (Richard St. Barbe Baker, *I Planted Trees*, 1944)[28]

Until the 19th century, most believed that ecosystems would always be able to recover from the pressure humanity had put on them. The fact that environmental pressures can push ecosystems' resilience past a threshold and into irreversible decline was understood and first articulated effectively to the mainstream in 1864 by George Perkins Marsh. Marsh emphasized, in his bestselling publication *Man and Nature: Or, Physical Geography as Modified by Human Action*, that some acts of destruction exceeded the Earth's recuperative powers:

> The ravages committed by man subvert the relations and destroy the balance which nature had established between her organized and her inorganic creations; and she avenges herself upon the intruder, by letting loose upon her defaced provinces destructive energies hitherto kept in check by organic forces destined to be his best auxiliaries, but which he has unwisely dispersed and driven from the field of action. When the forest is gone, the great reservoir of moisture stored up in its vegetable mould is evaporated, and returns only in deluges of rain to wash away the parched dust into which that mould has been converted ... The Earth is fast becoming an unfit home for its noblest inhabitant, and another era of equal human crime and human improvidence ... would reduce it to such a condition of impoverished productiveness, of shattered surface, of climatic excess, as to threaten the depravation, barbarism and perhaps even extinction of the species.[29]

Marsh was a senior US diplomat and his book *Man and Nature* was a bestseller and a very influential book in the late 19th century. Until the publication of *Man and Nature*, many had believed that it is always possible to pull back once humanity's environmental pressure starts to cause serious ecological collapse. However, often by then the ecosystem may have already passed the ecological threshold, and the collapse is either irreversible or the environmental pressure (pollution or other system change) will need to be reduced significantly (by 90 per cent or more) to allow the ecosystem to recover. This phenomenon is known as hysteresis.

However, the 2005 UN *Millennium Ecosystem Assessment* provides significant evidence that environmental pressures can push an ecosystem's resilience past a threshold and into irreversible decline. One of the examples featured in the UN Millennium Ecosystem Assessment was the collapse of the Newfoundland cod fishery (see Figure 2.8). This sudden collapse forced the indefinite closure of the fishery to commercial fishing in 2003. Until the late 1950s, the fishery was exploited by both migratory seasonal fleets and local fishermen. But from the late 1950s, offshore deep trawlers began exploiting the deeper part of the stock in larger quantities, leading to a large catch increase. Internationally agreed quotas in the early 1970s and, following the declaration by Canada of an Exclusive Fishing Zone in 1977, national quota schemes, ultimately failed to arrest the decline and collapse of this fishery. The stock collapsed rapidly due to very low population levels in the 1980s and early 1990s. The fishery was closed indefinitely from 2003.

All over the world we are seeing ecosystems and their ecosystem services already collapsing, from Australia's bluefin tuna stocks to the wheat fields of Western Australia being overcome by salinity, to the algae blooms suffocating lakes in the northern hemisphere. There are now significant global efforts to better understand where these ecological limits and tipping points are.[31] How is it that so many ecosystems are close to collapse or have already collapsed? Simply stated, it comes down to the fact that humanity has

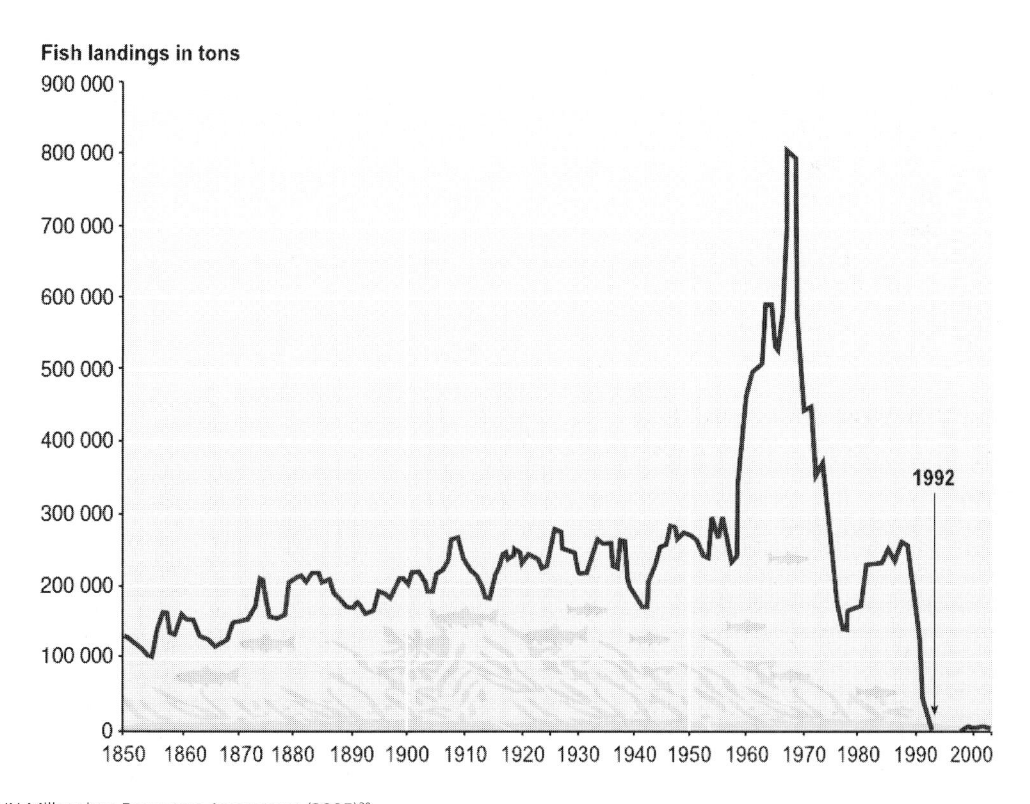

Source: UN Millennium Ecosystem Assessment (2005)[30]

Figure 2.8 *The collapsing of Atlantic cod stocks off the east coast of Newfoundland in 1992*

based its management of natural resources on flawed assumptions. Take the paradigm of maximum sustainable yield management of natural resources. In most cases the maximum sustainable yield is very close to the thresholds for collapse of that ecosystem. Also, in the past, there has been an expectation that change will be incremental and linear, when in fact with natural systems it is always non-linear. As reported in Chapter 2 of Hargroves and Smith's *The Natural Advantage of Nations*, rapid non-linear natural systems collapse, appropriately called 'environmental surprise', is occurring.[32]

Natural ecosystems are very complex. Therefore it is often hard to determine what a 'safe' level of pollutants is. It is also difficult to understand the causal links between pollutants and negative environmental effects – there is usually significant uncertainty. Faced with uncertainty, some often call for 'more research' to be done despite a history of scientists and health researchers warning in vain for decades about the dangers of many chemicals that were later recognized as pollutants. Examples include cigarettes and nicotine, asbestos[33] (first warning: 1898), PCBs[34] (1899), benzene[35] (1897) and acid rain[36] (1872), with causal links having been demonstrated between each of these chemicals and significant negative health and environmental health consequences. One of the reasons that causal links are hard to prove is that there is inherent uncertainty in natural systems, because the systems are so complex. Hence it often takes years and many people to collate enough data and analyse it to reduce the uncertainty significantly and to demonstrate a causal link. There is a long history of scientists' warnings being ignored about a range of issues due to such uncertainties, stemming from the complexity of natural systems and human health.

One of the reasons for the collapse of the Newfoundland cod fishery shown in Figure 2.8 is the significant uncertainty in assessing fish stocks.

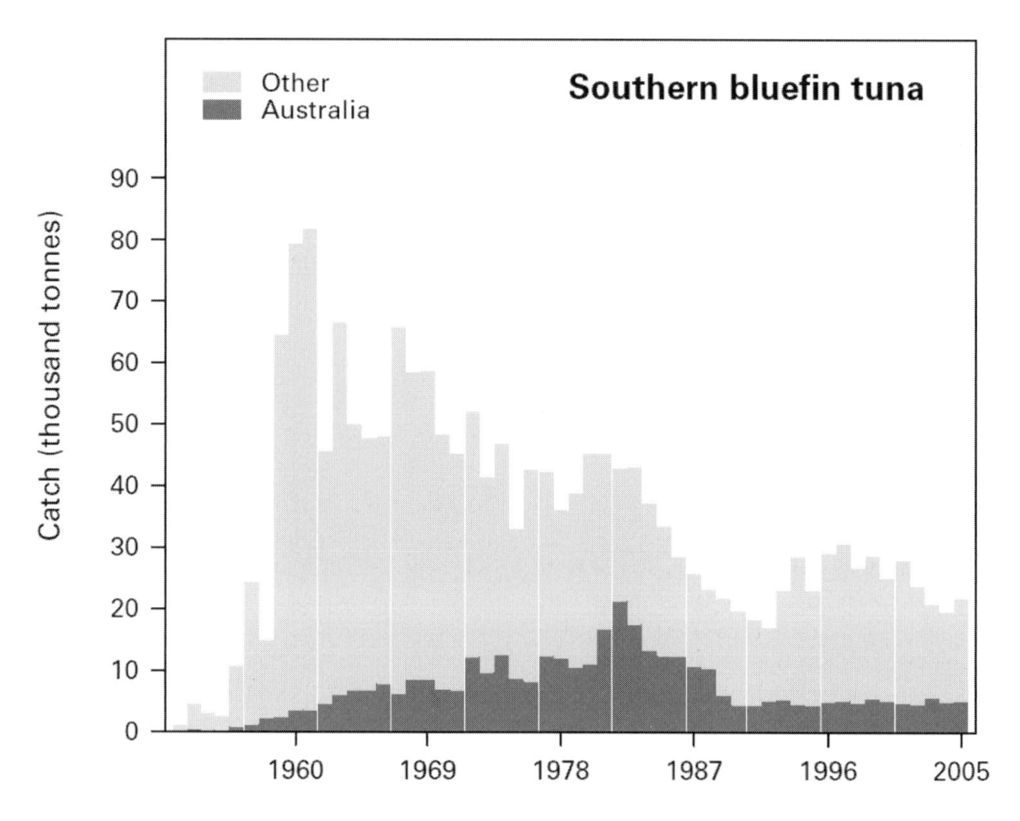

Source: Larcombe and McLoughlin (2006)[37]

Figure 2.9 *Southern bluefin tuna catch in thousands of tons, 1950–2006*

Government estimates of the state of fish stocks usually rely on the catch that fishermen report. It is too expensive and too difficult for governments to themselves go out into the oceans and take enough samples to know what the state of fish stocks are. Hence often by the time scientific consensus is built on an issue, it is decades after the concerns were raised by the original scientists. The catch history of the southern bluefin tuna shown in Figure 2.9 illustrates this.

By this time it is often too late and the ecological system is in irreversible decline or, at best, solving the problem will require a dramatic reduction of environmental pressures for the ecosystem in question to have a chance to recover.

Natural systems case study: Climate change

Addressing climate change in order to ensure that positive feedback loops in the Earth's biosphere are not unleashed is one challenge that will require a dramatic reduction of environmental pressures. The Intergovernmental Panel on Climate Change (IPCC) has warned that deep cuts to greenhouse gas emissions, of at least 60 per cent by 2050, will be needed to avoid dangerous climate change.[38] The Earth has a number of positive feedback loops that are already accelerating climate change. These are as follows.

Widespread melting of icebergs and ice-sheets

Already sea ice in the Arctic has shrunk to the smallest area ever recorded (see Figure 2.10).[39] Almost all the world's glaciers are now retreating. Ice has a high albedo effect, so it reflects heat, while water absorbs more heat, helping to warm the Earth faster and leading to more ice melting.

Permafrost

Permafrost, a permanently frozen layer of soil beneath the Earth's surface, is melting, releasing methane into the atmosphere (see Figure 2.11). The Western Siberia bog alone, which began melting in 2005, is believed to contain 70 billion tons of the gas.[41] Western Siberia has warmed faster than almost anywhere else on the planet, with an increase in average temperatures of some 3°C in the last 40 years.[42] The National Centre for Atmospheric Research estimates that 90 per cent of the top 10 feet of permafrost throughout the Arctic could thaw by 2100.

(a)

(b)

Source: NASA[40]

Figure 2.10 *The melting of the polar ice cap from (a) 1979 to (b) 2005*

Climate scientists now warn that it is critical to reduce greenhouse gas emissions rapidly to avoid passing a two degree warming threshold. The reason, they say, is because a two degree rise may invoke some additional critical reinforcing positive feedbacks. For instance, scientists predict that the terrestrial carbon sink (forest ecosystems and soils which are currently net sinks of CO_2) will change from net sinks to net sources of carbon. Another significant climate feedback mechanism is the Great Ocean Conveyor. Empirical studies show that rapid climate change has occurred in Earth's history when global warming has triggered the slowing and eventual halt of this significant ocean current that warms Europe (see Figure 2.12).[43]

If this ocean current slowed significantly or halted, the effects on human civilization would be devastating.

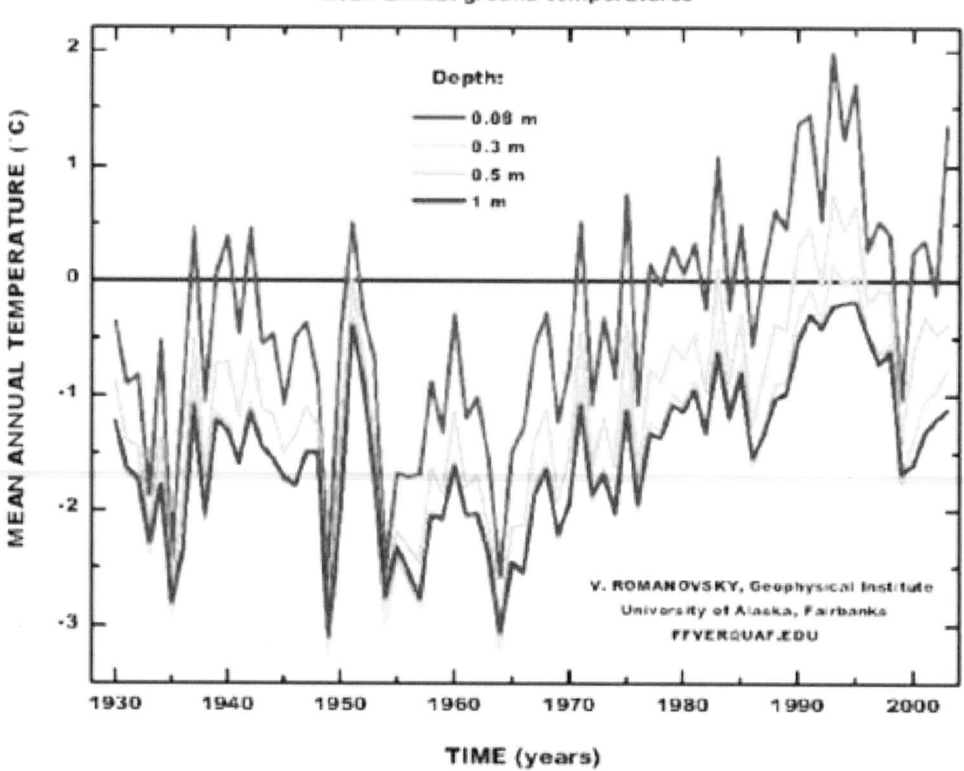

Source: Adapted from Arctic Climate Impact Assessment (2005)[44]

Figure 2.11 *Average annual ground temperature from Fairbanks, illustrating the warming trend observed across the Arctic that is causing permafrost to melt*

In addition, a negative feedback – global dimming – is being lessened by effective reductions in NOx, SOx and soot particulate emissions.[45] The first IPCC report in 1990 summed up why these reinforcing feedbacks are such a concern:[46]

> It appears likely that, as climate warms, these feedbacks will lead to an overall increase, rather than decrease, in natural greenhouse gas abundances. For this reason, climate change is likely to be greater than the estimates we have given.

Atmospheric carbon-dioxide levels that would increase climate change and unleash the positive feedbacks uncontrollably could possibly be reached in the coming decades. As James Hensen from NASA explains:[47]

> We live on a planet whose climate is dominated by positive feedbacks, which are capable of taking us to dramatically different conditions. The problem that we face now is that many feedbacks that came into play slowly in the past, driven by slowly changing forcings, will come into play rapidly now, at the pace of our human-made forcings, tempered a few decades by the oceans' thermal response time.

The risks of unleashing further positive feedbacks are well summarized by the 2006 UK Stern Review (see Figure 2.13). To ensure that humanity avoids the critical threshold of two degrees will require us to de-carbonize and transform the entire global industrial economy.

A new industrial revolution is needed which will be every bit as profound as the first industrial revolution.

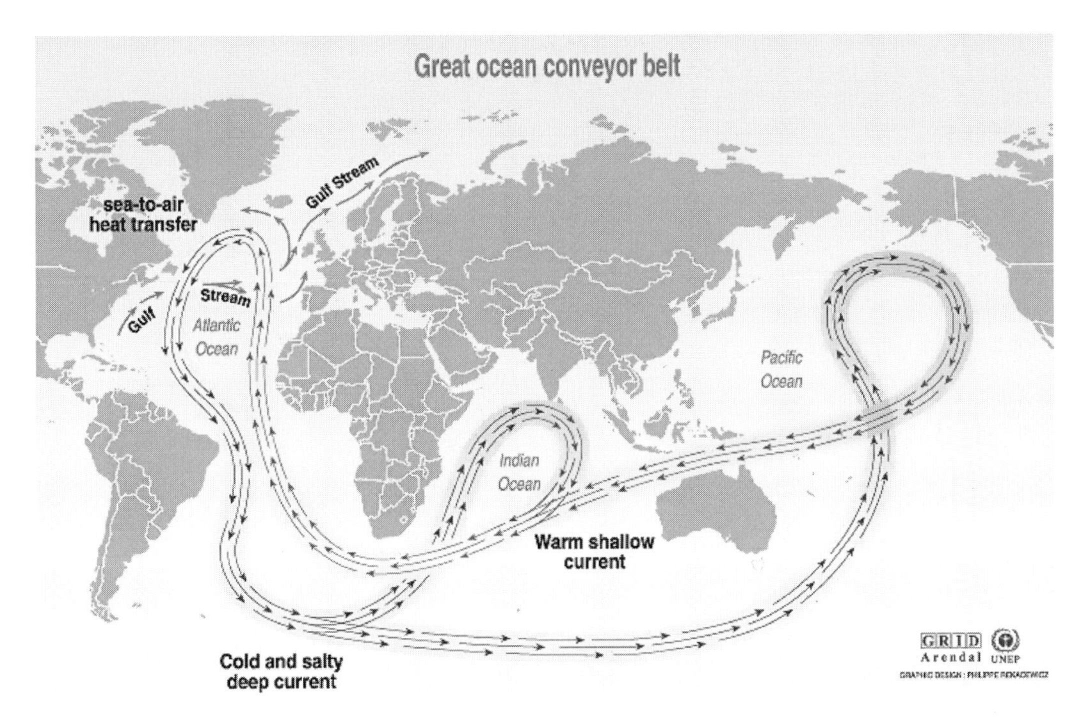

Source: UNEP (2007)[48]

Figure 2.12 *The Great Ocean Conveyor*

Former US Vice-President, Al Gore, in a recent address to US engineers at the US Embedded Systems Conference, argued that engineers can lead their societies in addressing climate change. Since there are significant energy efficiency opportunities of 30–60 per cent in most sectors of the economy and half of greenhouse gas emissions come from the built environment and infrastructure, engineers are in a very powerful position to make a positive difference.

As we will show in Chapters 4–10, advanced energy-efficiency strategies through WSD allow significant improvements in energy efficiency and reductions in greenhouse gas emissions. Engineers' ability to redesign technical systems to reduce significant greenhouse gas emissions on the planet is vital to preventing more positive feedbacks that may further destabilize the Earth's climate system. As Gore stated:[49]

> Those in the (technical) embedded-systems field can be a big part of a solution to the climate crisis. ... Embedded systems can be a big part of this.

Without going into the technical detail, Gore pointed to how 'power conservation and better efficiency are aids to lowering the amount of CO_2 released into the atmosphere. Asking better questions and systems design are really key to this.' He concluded that, 'An engineer is someone who has a vision and puts that vision into a solution. ... Engineers can lead this evolution, because engineering is making vision real.'[50]

As well as understanding more about natural systems, engineers also need to understand systems science, because many ideas of Systems Engineering and Whole System Approaches to Sustainable Design have been taken from advances in systems science. Hence we consider this next.

Science and systems science

Earlier in this chapter we outlined how, since many engineering systems have become more complex and engineers have become more specialized, there has emerged a recognized need for a new holistic, integrating discipline of engineering – Systems Engineering. A similar process has occurred in science. Over the last 200 years, the amount of scientific knowledge has expanded exponentially. This made it necessary to classify what was discovered into scientific

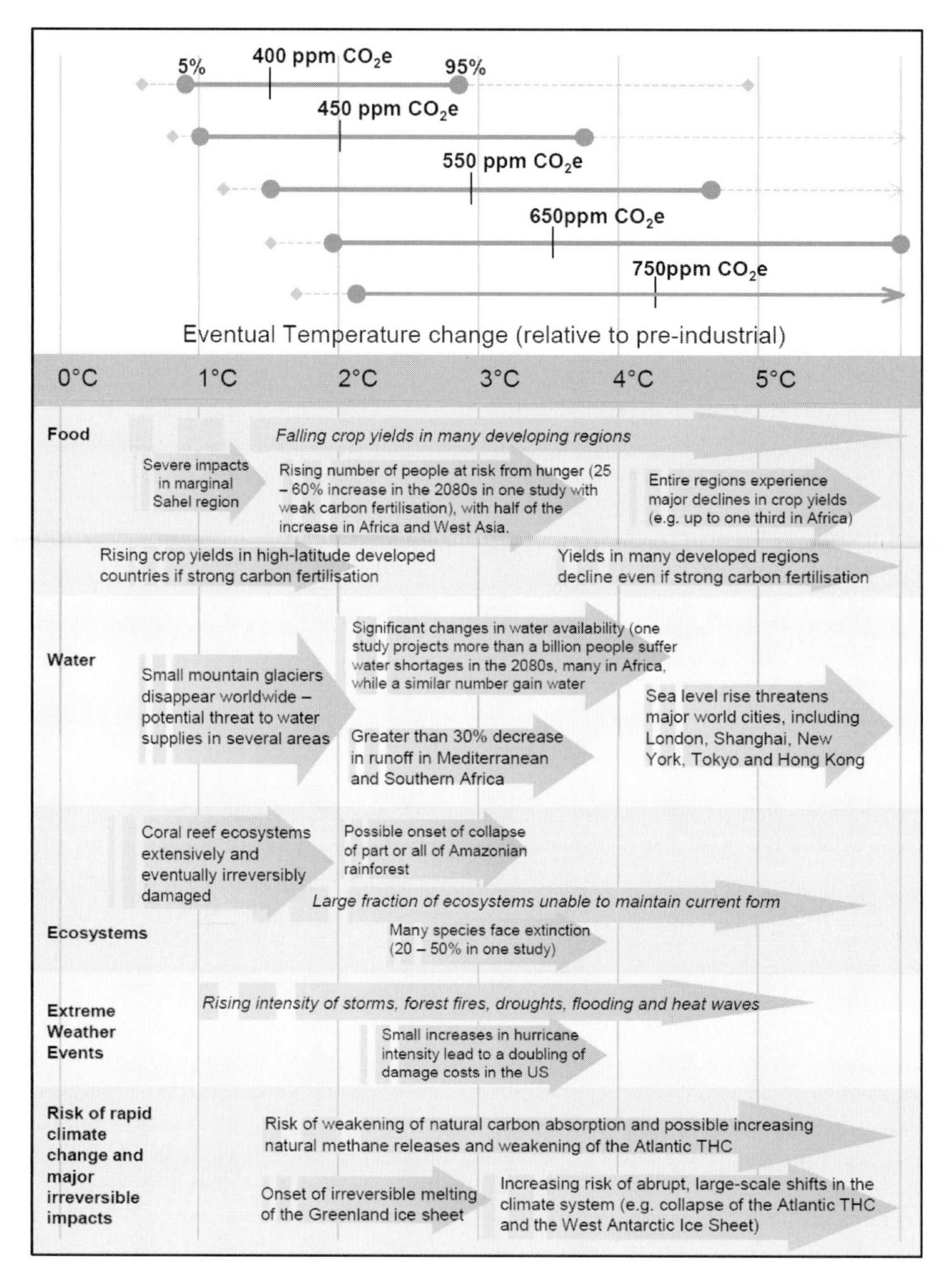

Source: Stern (2006)[51]

Figure 2.13 *Stabilization levels and probability ranges for temperature increases*

disciplines. Over the last two centuries, over 100 new scientific disciplines have been created to manage and classify this explosion of knowledge and discovery. Systems science is a relatively new unifying development based on the insight that systems have general characteristics, independent of the area of science to which they belong. Some key ideas of systems science are now discussed to show how these ideas are being used by Systems Engineering and WSD to help achieve sustainability. This is covered next by overviewing some key developments in cybernetics, general systems theory and systemology.

Cybernetics

The word 'cybernetics', first used in 1947 by Norbert Wiener, is from the Greek word for 'steersman'. Cybernetics is concerned with the role feedback plays in facilitating self-regulation of systems, whether mechanical, electrical, electromechanical or biological. Systems Engineering and WSD are focused on achieving a stated goal. To ensure this goal is achieved, often it is important to control dynamic processes using automated feedback. Engineering courses tend to focus on the aspects of control engineering and the role feedbacks can play to assist engineers to create 'cool' and complicated engineering systems such as robotics. But control engineering has a critical role to play throughout all of society to help reduce energy, water and materials waste and help achieve sustainability. An everyday example of this is the thermostat in domestic heaters. A thermometer measures temperature, allowing the user to program the heater to only come on at certain temperatures or at certain times of the day. This ensures that energy wastage is minimized. Numerous countries are now rolling-out smart metering to provide residential households with feedback every half hour on the amount of energy they are using to help them reduce their energy consumption. Control engineering is also critical in better managing distributed energy and waste systems that will be critical to achieving a cost-effective transition to a sustainable society. Engineers are already building automated feedback into many industrial processes to better manage these processes in real time and thus minimize the amount of energy, water and chemicals used. But there is still significant potential for engineers to use automated feedback more widely in order to help reduce energy and material usage and help achieve sustainability.

General Systems Theory

General Systems Theory is a broader unifying approach than cybernetics and was invented in the late 1940s by L. von Bertalanffy.[52] It is based on the premise that there are basic principles common to all systems and has gone well beyond the concept of control and automated self-regulation that is at the heart of cybernetics. The goal of General Systems Theory is to develop a framework for describing general relationships in the natural and human-made world. The goal is motivated by a desire to develop a common language and robust framework to facilitate communication and collaboration across the disciplines of science, social sciences and engineering. Blanchard and Fabrycky explain that:[53]

> One approach to creating such a framework is the structuring of a hierarchy of levels of complexity for simple units of behaviour in the various fields of research. A hierarchy of levels can lead to a systematic approach to systems that has broad application.

Kenneth Boulding's[54] efforts to do this are summarized in Table 2.1.

Another classification of General Systems Theory uses three organizing principles to define characteristics of systems: *rate of change*, *purpose* and *connectivity*. Each principle comprises a pair of 'polar-opposite' systems properties:[55]

1 Rate of change: structural (static) or functional (dynamic);
2 Purpose: purposive or non-purposive; and
3 Connectivity: mechanistic (mechanical) or organismic.

There are eight ways that these systems properties can be arranged to form eight general 'cells', or types of systems (Table 2.2).

Significant work has been done in the area of General Systems Theory to also analyse systems archetypes, in other words common system inter-relationships and patterns of behaviour that arise again and again in the real world.[56] These are well summarized in Peter Senge's classic text on systems theory, *The Fifth Discipline*.[57] Systems scientists have analysed many systems and developed systems archetypes to describe various standard types of common system relationships that arise again and again in the real world. Some forms of systems have common trends of behaviour and can be generally identified as being of a particular family, or 'archetype', as described in Table 2.3. Quite often one particular archetype may not fit a certain type of situation; hence it is possible to overlap a number of archetypes to more accurately describe system behaviour.

Table 2.1 *Kenneth Boulding's classification of systems*

Level	Characteristic	Examples	Relevant disciplines
1. Structures, frameworks	Static	Bridges	Description, verbal or pictorial, in any discipline
2. Dynamic system of clock-works	Predetermined motion (may exhibit equilibrium)	Natural physical universe	Chemistry, physics, natural sciences
3. Thermostat or cybernetic system	Closed-loop control	Thermostats	Control theory, cybernetics
4. The Level of the Cell, or open systems such as the cell, where life begins to be evident	Structurally self-maintaining	Biological cells	Theory of metabolism (information theory)
5. The Level of the Plant, with the genetic–societal structure making up the world of botany	Organized whole with functional parts, 'blue-printed', growth, reproduction	Plants	Botany
6. The Level of the Animal, encompassing mobility and self-awareness	A brain to guide total behaviour, ability to learn.	Birds	Zoology
7. The Level of the Human, encompassing self-consciousness	Self-consciousness, knowledge of knowledge, symbolic language.	Humans	Biology, psychology
8. Level of Social Organization	Roles, communication, transmission of values	Families	History, sociology, anthropology, behavioural science
9. The Level of Unknowables – transcendental systems	'Inescapable unknowables'	God	?

Source: Checkland (1999), p105[58]

Table 2.2 *Classification of systems according to Jordan's Principles*

Cell	Example
Structural – Purposive, Mechanical	A road network
Structural – Purposive, Organismic	A suspension bridge
Structural – Non-purposive, Mechanical	A mountain range
Structural – Non-purposive, Organismic	A bubble (or any physical system in equilibrium)
Functional – Purposive, Mechanical	A production line (a breakdown in one machine does not affect other machines)
Functional – Purposive, Organismic	Living organisms
Functional – Non-purposive, Mechanical	The changing flow of water as a result of a change in the river bed
Functional – Non-purposive, Organismic	The space–time continuum

Source: Checkland (1999), p105[59]

Conclusion: Transition to the systems age

> The world – including the physical ecosystems and the societies that exist within them – is facing a new set of problems, the scale and complexity of which are unmatched in human history... We are currently stuck in a re-enforcing cycle. Our entrenched life-style and mind-set will continue to lead to unsustainable results unless these issues are addressed at the source... As problems associated with societal design arise (finite oil supply), governments are attempting to choose 'winners' for energy technologies while not considering impacts on the larger system. In so doing, there is the very real potential to create more problems than they solve (take the unsustainability of certain biofuels)... These complex problems require a whole-system approach in order to find long-term solutions that address the root causes of these impacts. (Archie Kasnet, Greenland Enterprises, 2008)[60]

Kasnet's quote is indicative of a significant shift that comes out of the recognition that taking a systems approach is more effective in addressing today's challenges.

Table 2.3 *Systems archetypes*

Systems archetype	Behaviour	Example
Reinforcing loop: an important variable accelerates up/down, with exponential growth/collapse.	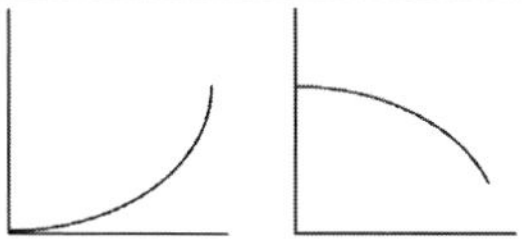	Climate change melts ice, reducing the albedo effect, further warming the planet.
Balancing loop: oscillating around a single target (with delay), or movement towards a target (without delay).	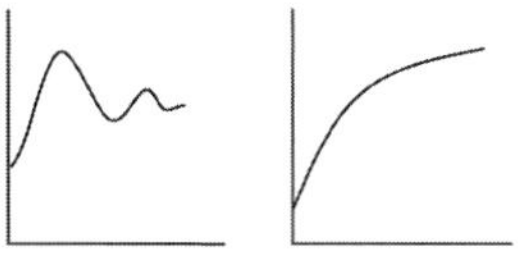	Managing population levels of an endangered species.
'*Fixes that backfire*': a problem symptom temporarily improves and then deteriorates, worse than before.	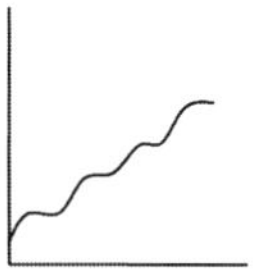	Negative rebound effects from efficiency investments.
'*Limits to growth*': there is growth (sometimes dramatic), then falling into decline or levelling off.	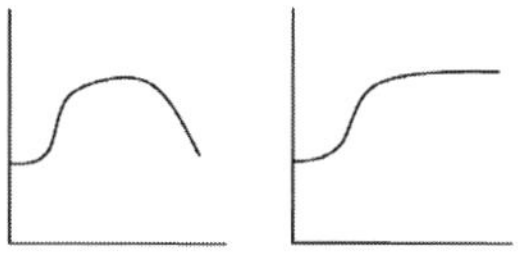	Oil production rates have peaked and are now in decline in over 60 countries.
'*Shifting the burden*': three patterns exist – reliance on a short-term fix grows, while efforts to fundamentally correct the real problem decline, and the problem symptom alternately improves and deteriorates.	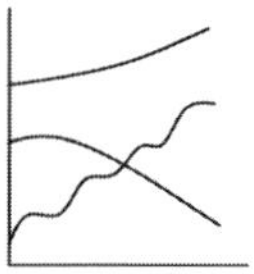	Modern agriculture's dependence on artificial fertilizers leads to algae blooms downstream.
'*Tragedy of the commons*': total activity grows, but gains from individual activities decline.	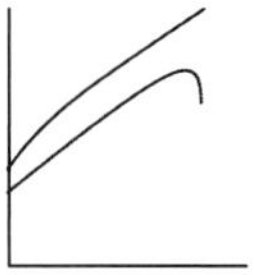	Collapse of fishing stocks.
'*Accidental adversaries*': each competitor's performance stays low or declines, while hostility increases over time.	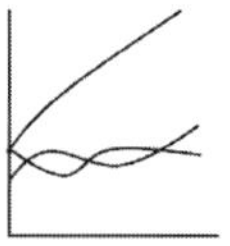	Disputes between supplier and manufacturer.

Source: Senge et al (1998)[61]; examples added by The Natural Edge Project

Over the last 200 years, humanity has sought to achieve progress through a largely reductionist approach to technological innovation and problem-solving. The reductionist approach has been very successful and has helped advance society. But the world we live in today is very different to the world 200 years ago. Through the advent of advanced technologies in communication and transportation, time barriers have dramatically compressed. Every aspect of human existence has become more inter-related and intertwined, with an increasingly complex set of relationships due to the phenomena of globalization – which has been particularly enhanced by the uptake of access to the internet. At the same time, better global communications are raising expectations of consumers in the Third World, who now aspire to First World living standards. Increasing global population and the desire for larger and better systems is leading to greater and greater levels of resource exploitation and environmental degradation. This means, for example, that the levels of greenhouse gas pollution in countries far from Australia now can play a part in bleaching coral reefs there. There is a growing recognition that as society progresses and becomes more technologically complex, the large-scale environmental problems that we face can only be addressed effectively through an integrated systems approach. As Blanchard and Fabrycky argue:[62]

> There is considerable evidence to suggest that the advanced nations of the world are leaving one technological age and entering another. ... It appears that this transition is bringing about a change in the conception of the world in which we live. This conception is both a realization of the complexity of natural and human-made systems and a basis for the improvement in people's position relative to these systems. ... Although eras do not have precise beginnings, the 1940s can be said to have contained the beginning of the end of the Machine Age and the beginning of the Systems Age.

In this new century we need to combine the best of reductionist knowledge, Systems Engineering and enhance this with a Whole System Approach to Sustainable Design. Chapters 3–5 next consider how to enhance conventional Systems Engineering through a Whole System Approach to Sustainable Design. They outline the key operational steps and processes of a Whole System Approach to Sustainable Design informed by the fundamentals of Systems Engineering. This new Whole System Approach to Sustainable Design, outlined next in Chapter 3, will help engineers proactively reduce the environmental, social and economic risks of their design projects. Chapters 3–5 are designed to create a robust framework to then, as Al Gore stated above, ask better questions in the technical worked examples in Chapters 6–10 to achieve a Whole System Approach to Sustainable Design.

Optional reading

Benyus, J. (1997) *Biomimicry: Innovation Inspired by Nature*, HarperCollins, New York

Birkeland, J. (ed) (2002) *Design for Sustainability: A Sourcebook of Ecological Design Solutions*, Earthscan, London

Department of the Environment and Heritage (2001) *Product Innovation: The Green Advantage: An Introduction to Design for Environment for Australian Business*, DEWR, www.environment.gov.au/settlements/industry/finance/publications/producer.html, accessed 5 January 2007

Hawken, P., Lovins, A. B. and Lovins, L. H. (1999) *Natural Capitalism: Creating the Next Industrial Revolution*, Earthscan, London, www.natcap.org, accessed 5 January 2007

Lyle, J. (1999) *Design for Human Ecosystems*, Island Press, Washington, DC

McDonough, W. and Braungart, M. (2002) *Cradle to Cradle: Remaking the Way We Make Things*, North Point Press, New York

Pears, A. (2004) 'Energy efficiency – Its potential: Some perspectives and experiences', background paper for International Energy Agency Energy Efficiency Workshop, Paris, April

Van der Ryn, S. and Calthorpe, P. (1986) *Sustainable Communities: A New Design Synthesis for Cities, Suburbs and Towns*, Sierra Club Books, San Francisco, CA

Von Weizsäcker, E., Lovins, A. and Lovins, L. (1997) *Factor Four: Doubling Wealth, Halving Resource Use*, Earthscan, London, www.wupperinst.org/FactorFour/index.html, accessed 5 January 2007

Notes

1 Rocky Mountain Institute (1997) 'Tunnelling through the cost barrier', *RMI Newsletter*, summer 1997, pp1–4, www.rmi.org/images/other/Newsletter/NLRMIsum97.pdf, accessed 5 January 2007.

2 Anderson, D. M. (P.E., fASME, CMC) (2008) *Design for Manufacturability and Concurrent Engineering, How to Design for Low Cost, Design in High Quality, Design for Lean Manufacture, and Design Quickly for Fast Production*, CIM Press, www.halfcostproducts.com/dfm_article.htm, accessed 11 July 2007.

3 Anderson, D. M. (P.E., fASME, CMC) (2008) *Design for Manufacturability and Concurrent Engineering, How to Design for Low Cost, Design in High Quality, Design for Lean Manufacture, and Design Quickly for Fast Production*, CIM Press, www.halfcostproducts.com/dfm_article.htm, accessed 11 July 2007.

4 Honour, E. C. (2004) *Understanding the Value of Systems Engineering*, proceedings of the Fourteenth Annual Symposium of the International Council on Systems Engineering, Toulouse, France, www.incose.org/secoe/ 0103/ValueSE-INCOSE04.pdf, accessed 16 July 2007.

5 Reprinted with permission. Original source: Professor Paul G. Ranky 'Concurrent engineering and PLM (Product Lifecycle Management)', published 2002–2008 by CIMware USA, Inc., www.cimwareukandusa.com, www.cimwareukandusa.com/All_IE655/IE655Spring2007.html, accessed 16 July 2007.

6 Honour, E. C. (2004) *Understanding the Value of Systems Engineering*, proceedings of the Fourteenth Annual Symposium of the International Council on Systems Engineering, Toulouse, France, www.incose.org/secoe/0103/ValueSE-INCOSE04.pdf, accessed 16 July 2007.
7 This refers to the Victorian era of the 19th century.
8 Blanchard, B. S. and Fabrycky, W. J. (2006) *Systems Engineering and Analysis* (fourth edition), Pearson Prentice Hall, Upper Saddle River, NJ.
9 Blanchard, B. S. and Fabrycky, W. J. (2006) *Systems Engineering and Analysis* (fourth edition), Pearson Prentice Hall, Upper Saddle River, NJ, Chapter 1.
10 Adcock, R. (n.d.) 'Principles and practices of systems engineering', presentation, Cranfield University, p8, www.incose.org.uk/Downloads/AA01.1.4_Principles%20&%20practices%20of%20SE.pdf, accessed 27 March 2008. This figure is based on text in Flood, R. L. and Carson, E. R. (1993) *Dealing with Complexity: An Introduction to the Theory and Application of Systems Science* (second edition), Plenum Press, New York, p17.
11 Blanchard, B. S. and Fabrycky, W. J. (2006) *Systems Engineering and Analysis* (fourth edition), Pearson Prentice Hall, Upper Saddle River, NJ, Chapter 1.
12 Blanchard, B. S. and Fabrycky, W. J. (2006) *Systems Engineering and Analysis* (fourth edition), Pearson Prentice Hall, Upper Saddle River, NJ, p45.
13 Blanchard, B. S. and Fabrycky, W. J. (2006) *Systems Engineering and Analysis* (fourth edition), Pearson Prentice Hall, Upper Saddle River, NJ, p45.
14 Hitchins, D. K. (2003) *Advanced Systems Thinking, Engineering and Management*, Artech House, Norwood, MA.
15 Lewis, J. (1985) 'Lead poisoning: A historical perspective', *EPA Journal*, May, www.epa.gov/history/topics/perspect/lead.htm, accessed 5 January 2007.
16 Elkins, J. (1999) 'Chlorofluorocarbons (CFCs)', in D. E. Alexander and R. W. Fairbridge (1999) *The Chapman and Hall Encyclopaedia of Environmental Science*, Kluwer Academic, Boston, MA, pp78–80, www.cmdl.noaa.gov/noah/publictn/elkins/cfcs.html, accessed 5 January 2007.
17 Commoner, B. (1972) *The Closing Circle: Nature, Man and Technology*, Bantam Books, Toronto, Canada.
18 Hawken, P., Lovins, A. B. and Lovins, L. H. (1999) *Natural Capitalism: Creating the Next Industrial Revolution*, Earthscan, London, Chapter 14: 'Human capitalism', www.natcap.org/images/other/NCchapter14.pdf, accessed 13 August 2007.
19 Adcock, R. (n.d.) 'Principles and practices of systems engineering', presentation, Cranfield University, p6, www.incose.org.uk/Downloads/AA01.1.4_Principles%20&%20practices%20of%20SE.pdf, accessed 27 March 2008. This figure is based on a figure in Flood, R. L. (1987) cited in Flood, R. L. and Carson, E. R. (1993) *Dealing with Complexity: An Introduction to the Theory and Application of Systems Science* (second edition), Plenum Press, New York, p74.
20 Blanchard, B. S. and Fabrycky, W. J. (2006) *Systems Engineering and Analysis* (fourth edition), Pearson Prentice Hall, Upper Saddle River, NJ, Chapter 1.
21 Proust, K., Dovers, S., Foran, B., Newell, B., Steffen, W. and Troy, P. (2007) *Climate, Energy and Water: Accounting for the Links*, Land and Water Australia, Canberra, www.lwa.gov.au/downloads/publications_pdf/ER071256.pdf, accessed 19 October 2007.
22 Tenner, E. (1997) *Why Things Bite Back*, Fourth Estate, London; Sterman, J. D. (2000) *Business Dynamics: Systems Thinking and Modeling for a Complex World*, Irwin McGraw-Hill, Boston, MA; Jervis, R. (1997) *System Effects: Complexity in Political and Social Life*, Princeton University Press, Princeton, NJ; Newell, B. and Proust, K. (2004) *The Darwin Harbour Modelling Project: A Report to the Ecological Research Group of the Darwin Harbour Advisory Committee*, Darwin Harbour Advisory Committee, Darwin, Australia, www.nt.gov.au/nreta/water/dhac/publications/pdf/finalreport20050307.pdf, accessed 19 October 2007; Proust, K. and Newell, B. (2006) *Catchment and Community: Towards a Management-Focused Dynamical Study of the ACT Water System*, ACTEW Corporation, Canberra, www.water.anu.edu.au/pdf/publications/Catchment%20and%20Community.pdf, accessed 19 October 2007.
23 Holtzclaw, J. (n.d.) *Stop Sprawl: Induced Traffic Confirmed*, Sierra Club, San Francisco, CA, US, www.sierraclub.org/sprawl/transportation/seven.asp, accessed 11 August 2008.
24 Acevedo, W. (1999) *Analyzing Land-use Change in Urban Environments*, USGS Fact Sheet 188–99, US Geological Survey, http://landcover.usgs.gov/urban/info/factsht.pdf, accessed 11 August 2008.
25 Schneider, A. and NASA Landsat cited in NASA Goddard Space Flight Centre, *2003 Earth Feature Story*, NASA Goddard Space Flight Centre, www.gsfc .nasa.gov/feature/2003/1212globalcities.html, accessed 18 March 2008.
26 See UN Millennium Ecosystem Assessment at www.maweb.org/en/index.aspx, accessed 28 March 2008
27 Diamond, J. (2005) *Collapse: How Societies Choose to Fail or Succeed*, Viking, New York.
28 St. Barbe Baker, R. (1944) *I Planted Trees*, Lutterworth Press, London.
29 Marsh, G. P. (1864) *Man and Nature: Or, Physical Geography as Modified by Human Action*, Weyerhauser Environmental Classics Series, University of Washington Press, London
30 Millennium Ecosystem Assessment (2005) *Ecosystems and Human Well-being: General Synthesis*, Island Press,

Washington, DC, p12, www.millenniumassessment.org/en/Synthesis.aspx, accessed 18 March 2008.

31 See The Resilience Alliance – 'Resilience's thresholds database' – at www.resalliance.org/ev_en.php, accessed 2 July 2007.

32 Hargroves, K. and Smith, M. (eds) (2005) *The Natural Advantage of Nations*, Earthscan, London, Chapter 2, www.thenaturaladvantage.info, accessed 2 June 2007.

33 Deane, L. (1898) 'Report on the health of workers in asbestos and other dusty trades', in HM Chief Inspector of Factories and Workshops (1898) *Annual Report for 1898*, HMSO London, pp171–172. (see also the Annual Reports for 1899 and 1900).

34 Polychlorinated biphenyls (PCBs) are chlorinated organic compounds that were first synthesized in the laboratory in 1881. By 1899 a pathological condition named chloracne had been identified, a painful disfiguring skin disease that affected people employed in the chlorinated organic industry. Mass production of PCBs for commercial use started in 1929.

35 Santessen, C. G. (1897) 'Chronische Vergiftungen mit Steinkohlentheerbenzin: Vier Todesfalle' ['Chronic poisoning with Steinkohlentheerbenzin: Four death case'], *Arch. Hyg. Bakteriol*, vol 31, pp336–376.

36 Smith, R. A. (1872) *Air and Rain*, Longmans Green and Co., London.

37 Larcombe, J. and McLoughlin, K. (eds) (2006) *Fishery Status Report 2006: Status of Fish Stocks Managed by the Australian Government*, Bureau of Rural Sciences, p105, http://affashop.gov.au/product.asp?prodid=13736, accessed 18 March 2008.

38 IPCC (2001) *Climate Change 2001 Third Assessment Report: The Scientific Basis*, IPCC, Cambridge University Press, Cambridge, UK.

39 National Snow and Ice Data Centre (2005) 'Sea ice decline intensifies', press release, National Snow and Ice Data Centre, http://nsidc.org/news/press/20050928_trendscontinue.html, accessed 2 June 2007; the National Snow and Ice Data Center (NSIDC) is a part of the Cooperative Institute for Research in Environmental Sciences at the University of Colorado, Boulder, CO.

40 National Aeronautics and Space Administration (NASA) (2005) *Arctic Sea Ice Continues to Decline, Arctic Temperatures Continue to Rise In 2005*, NASA, www.nasa.gov/centers/goddard/news/topstory/2005/arcticice_decline.html, accessed 8 May 2008.

41 Pearce, F. (2005) 'Climate warning as Siberia melts', *New Scientist*, 11 August, http://environment.newscientist.com/channel/earth/mg18725124.500-climate-warning-as-siberia-melts.html, accessed 18 March 2008.

42 Pearce, F. (2005) 'Climate warning as Siberia melts', *New Scientist*, 11 August, http://environment.newscientist.com/channel/earth/mg18725124.500-climate-warning-as-siberia-melts.html, accessed 18 March 2008.

43 Bryden, H. L. et al (2005) 'Slowing of the Atlantic Meridional Overturning Circulation at 250N', *Nature*, vol 438, pp655–657.

44 Walsh, J. E., Anisimov, O., Hagen, J. O. M., Jakobsson, T., Oerlemans, J. Prowse, T. D., Romanovsky, V., Savelieva, N., Serreze, M., Shiklomanov, A., Shiklomanov, I. and Solomon, S., (2005) 'Cryosphere and hydrology', Chapter 6 in C. Symon, L. Arris and B. Heal (eds) *Arctic Climate Impact Assessment Scientific Report*, Cambridge University Press, p210, www.acia.uaf.edu/pages/scientific.html, accessed 15 March 2008.

45 Wild, W. et al (2005) 'From dimming to brightening: Decadal changes in solar radiation at Earth's surface', *Science*, vol 308, pp847–850.

46 Houghton, J. T., Jenkins, G. J. and Ephraums, J. J. (eds) (1990) *Climate Change: The IPCC Scientific Assessment*, IPCC, Cambridge University Press, Cambridge, UK.

47 Hansen, J. (2006) *Communicating Dangers and Opportunities in Global Warming*, Draft report, American Geophysical Union, San Francisco, CA.

48 UNEP (2007) *Potential Impacts of Climate Change*, UNEP, www.grida.no/climate/vital/impacts.htm, accessed 2 July 2007.

49 Deffree, S. (2007) 'Gore: Climate crisis could attract next generation of engineers', *Electronic News*, www.reed-electronics.com/semiconductor/article/ CA6430597?industryid=3140&nid=2012, accessed 2 July 2007.

50 Deffree, S. (2007) 'Gore: Climate crisis could attract next generation of engineers', *Electronic News*, www.reed-electronics.com/semiconductor/article/CA6430597?industryid=3140&nid=2012, accessed 2 July 2007.

51 Stern, N. (2006) *Stern Review: The Economics of Climate Change*, HM Treasury, UK, Chapter 13: 'Towards a goal for a climate', p294, Figure 13.4, www.hm-treasury.gov.uk/ independent_reviews/stern_review_economics_climate_change/sternreview_index.cfm, accessed 3 January 2007.

52 Bertalanffy, L. (1951) 'General system theory: A new approach to unity of science', *Human Biology*, vol 23, no 4, pp303–361.

53 Blanchard, B. S. and Fabrycky, W. J. (2006) *Systems Engineering and Analysis* (fourth edition), Pearson Prentice Hall, Upper Saddle River, NJ, Chapter 1.

54 Boulding, K. (1956) 'General systems theory: The skeleton of science', *Management Science*, vol 2, no 3, pp197–208, www.panarchy.org/boulding/systems.1956.html, accessed 24 September 2007.

55 Checkland, P. (1999) *Systems Thinking, Systems Practice*, John Wiley Publishing, Chichester, UK, pp107–108.

56 Senge, P. M., Kleiner, A., Roberts, C., Ross, R. B. and Smith, B. J. (1994) *The Fifth Discipline Fieldbook*, Nicholas Brealy Publishing, London, pp121–150.
57 Senge, P. M. et al (1998) *The Fifth Discipline Fieldbook*, Nicholas Brealy Publishing, London, pp121–150.
58 Checkland, P. (1999) *Systems Thinking, Systems Practice*, John Wiley Publishing, Chichester, UK.
59 Checkland, P. (1999) *Systems Thinking, Systems Practice*, John Wiley Publishing, Chichester, UK, pp107–108.
60 Archie Kasnet, Greenland Enterprises, personal communication on 29 September 2008.
61 Senge, P. M. et al (1998) *The Fifth Discipline Fieldbook*, Nicholas Brealy Publishing, London, pp121–150.
62 Blanchard, B. S. and Fabrycky, W. J. (2006) *Systems Engineering and Analysis* (fourth edition), Pearson Prentice Hall, Upper Saddle River, NJ, Chapter 1.

3

Enhancing the Systems Engineering Process through a Whole System Approach to Sustainable Design

Educational aim

Chapter 1 introduced the concept of a Whole System Approach to Sustainable Design and outlined some of the benefits of undertaking such a design process to help society achieve ecological sustainability. It featured exciting case studies of the work of leading Whole System Designers like Amory Lovins and the Rocky Mountain Institute. Chapter 2 showed that Systems Engineering is an established field of engineering that has been developed, like a Whole System Approach, to address the same weaknesses of traditional specialized engineering. This chapter illustrates clearly how a Whole System Approach fits into the traditional engineering methodologies of Systems Engineering that are taught in engineering schools all around the world. It outlines traditional operational Systems Engineering processes as described in leading Systems Engineering textbooks and highlights how they can be further enhanced through a Whole System Approach to Sustainable Design. Starting with an overview of the standard phases of Systems Engineering in practice that are common to most engineering projects and problem-solving exercises, the chapter demonstrates that there is a need to more explicitly include sustainability considerations in Systems Engineering. It then identifies 10 key operational elements of a Whole System Approach to Sustainable Design that enhance traditional Systems Engineering to assist all designers to leave a positive legacy. It includes two key diagrams that summarize the traditional Systems Engineering process (See Figure 3.1) and how it can be enhanced through a Whole System Approach (See Figure 3.3). If you are teaching or lecturing this chapter or using it for a tutorial, both figures are designed to be useful handouts for the class to help summarize the key points.

Required reading

Blanchard, B. S. and Fabrycky, W. J. (2006) *Systems Engineering and Analysis* (fourth edition), Pearson Prentice Hall, Upper Saddle River, NJ, pp3–6 and 17–32

Hawken, P., Lovins, A. B. and Lovins, L. H. (1999) *Natural Capitalism: Creating the Next Industrial Revolution*, Earthscan, London, Chapter 6, pp111–124, www.natcap.org/images/other/NCchapter6.pdf, accessed 3 October 2007

International Council on Systems Engineering (1993) 'An identification of pragmatic principles: Final report', International Council on Systems Engineering, pp7–10, www.incose.org/ProductsPubs/pdf/techdata/PITC/PrinciplesPragmaticDefoe_1993-0123_PrinWG.pdf, accessed 3 October 2007

Systems engineering in practice

The typical system life-cycle consists of development (including production), operation and retirement stages. Systems Engineering is performed primarily during the development stage, excluding production, while the production, operation and retirement stages are key design considerations.

The Systems Engineering process consists of several phases. The name, aims and activities of each phase vary between literature sources. Here, the Systems Engineering process is presented in four phases that are approximately consistent with the phases presented in popular literature sources. Phase 2, Phase 3 and Phase 4 feed back into earlier phases, so the process may be iterative in practice.

Phase 1: Need definition

The aim of the Need Definition phase is to develop an understanding of the system, its purpose and its feasibility. This phase typically involves:

- Performing a feasibility study, including requirements analysis and trade-off studies;
- Drafting design specifications using customer interaction, quality function deployment and benchmarking against best-in-class competitors; and
- Planning.

Phase 2: Conceptual design

The aim of the Conceptual Design phase is to thoroughly explore the solution space for all possible options that address the Need Definition; and then to generate a set of conceptual systems for further development. This phase typically involves:

- Researching;
- Performing a functional analysis and decomposition;
- Brainstorming a set of conceptual systems; and
- Short-listing the conceptual systems by testing against the draft design specifications.

Phase 3: Preliminary design

The aim of the Preliminary Design phase is to develop the set of conceptual systems into a set of preliminary systems; and then to select the best system for further development. This phase typically involves:

- Developing a physical architecture;
- Designing major physical subsystems and components while incorporating 'Design for X';
- Defining interfaces;
- Selecting the best preliminary system by testing against the draft design specifications; and
- Developing detailed design specifications.

Phase 4: Detail design

The aim of the Detail Design phase is to develop the selected preliminary system into the detail system. This phase typically involves:

- Designing physical subsystems and components in detail while incorporating production-based 'Design for X';
- Subsystem testing and refining; and
- Integration and system testing against the detailed design specifications.

Figure 3.1 shows the typical technical design activities and interactions in the four phases throughout the system life-cycle. Note that in the figure, some activities of the Conceptual Design phase are absorbed into the Preliminary Design phase. Figure 3.2 shows the typical hierarchy of design considerations that can guide design specifications. Lower-order considerations depend on higher-order considerations.

Potential modifications to the systems engineering process

Blanchard and Fabrycky loosely define Systems Engineering as 'good engineering with special areas of emphasis', which includes a top–down approach, a life-cycle orientation, a more complete definition of system requirements and an interdisciplinary team approach.[3] It is proposed that an additional overarching area of emphasis, sustainability, is required to promote the development of systems that are in balance with the Earth's capacity.

Incorporating sustainability considerations primarily affects two of the original areas of emphasis:

1. *Emphasize sustainable end-of-life options*: The original emphasis on life-cycle orientation suggests that decisions need to be based on life-cycle considerations, highlighting the often overlooked impact on the production, operation (including maintenance and support) and retirement (specified as disposal) stages. Incorporating a sustainability emphasis will primarily affect the considerations for the retirement stage, which originally emphasized disposal over other end-of-life options. In Figure 3.1, 'Phase-out and Disposal' becomes 'Retirement', for which activities include recovery, disassembly and prioritized end-of-life

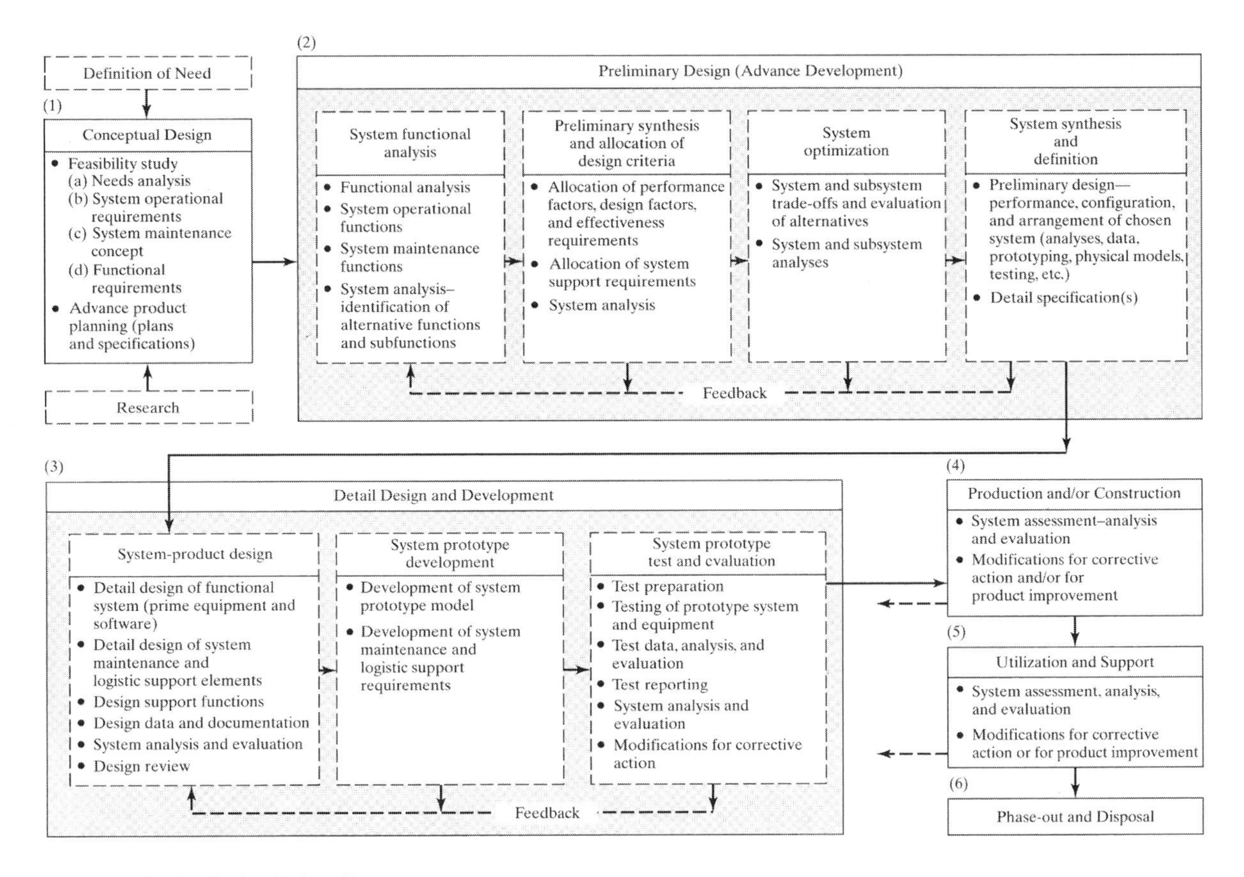

Source: Blanchard and Fabrycky (2006)[1]

Figure 3.1 *Systems Engineering technological design activities and interactions by phase*

processing – reuse, remanufacturing, recycling and disposal. In Figure 3.2, 'Retirement and Disposal Cost' becomes 'Retirement Cost'. Considerations are added to 'Recycle Cost', and considerations are added to 'Disposability'. These modifications will shift the emphasis from disposable systems and built-in obsolescence to closed-loop systems.

2 *Emphasize sustainable resource use (energy, material and water inputs and outputs)*: The original emphasis on a more complete definition of system requirements suggests that the true system requirements need to be identified and made traceable from the system level downward. Incorporating a sustainability emphasis will primarily affect the considerations for the production and operation stages, which originally did not emphasize resource use and did emphasize pollutability over restorability as the measure of biological impact. In Figure 3.2, considerations are added to 'Operating Cost' – 'Energy Consumption' is upgraded from a fifth-order consideration to a fourth-order consideration, third-order considerations are added to 'System Effectiveness', and considerations are added to 'Pollutability'. These modifications will place an emphasis on resource productivity, Factor X targets, and benign and restorative design.

These two key features are also largely absent in other key Systems Engineering texts. Ulrich and Eppinger, authors of *Product Design and Development*[4] (another Systems Engineering textbook used in university undergraduate engineering courses), do not emphasize sustainable end-of-life options or sustainable resource use. Dieter,[5] author of *Engineering Design: A Materials and Processing Approach* (another Engineering Design textbook used in university undergraduate engineering courses), discusses recycling as an end-of-life alternative

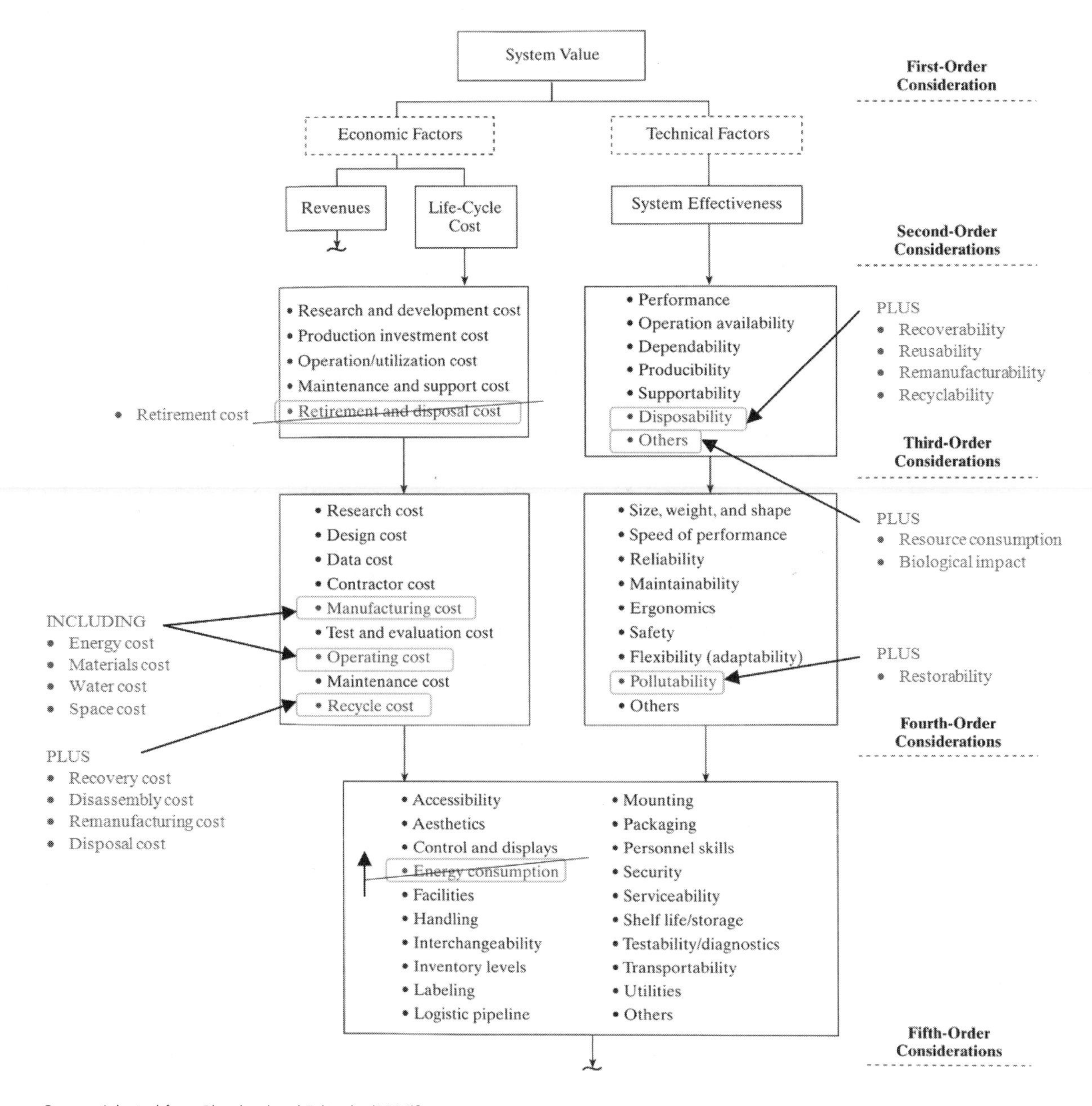

Source: Adapted from Blanchard and Fabrycky (2006)[2]

Figure 3.2 *Hierarchy of design considerations*

to disposal and briefly mentions a few aspects of sustainable resource use in a review of 'design for environment'. The International Council on Systems Engineering[6] emphasizes disposal over sustainable end-of-life options and does not emphasize sustainable resource use.

Incorporating sustainability into the process more explicitly

A Whole System Approach builds on from Systems Engineering by more explicitly emphasizing the steps required to develop sustainable systems.

Sustainability is emphasized through the following activities:

- Sustainability considerations are brought to the fore along with economic and performance considerations. The main sustainability considerations are resource use (energy, material and water inputs and outputs), biological impact and providing options for future generations. These considerations are incorporated in the specifications during the Need Definition phase to ensure they are regularly consulted during the remaining phases. They also ensure that benchmarking is against a sustainable system, not just the best-in-class system.
- Research is emphasized as an early step during the Conceptual Design, Preliminary Design and Detail Design phases. Research in each phase is used to populate a database of possible technological and design options of suitable scope for the phase. Latest technological innovations can provide opportunities for effectively fulfilling the specifications without compromising on some aspect of performance.
- There are elements that help streamline the design process during the Preliminary Design phase and the optimization process during the Detail Design phase. Streamlining these processes shortens the time required to converge on the optimal solution. The elements encourage developing the global optimal system of the entire solution space. Ignoring these elements generally encourages developing, at best, the local optimal system, that is, a system that is optimal given arbitrary constraints imposed by some ill-selected subsystem or component. The elements are based on the Whole System Design (WSD) precepts presented in the book *Natural Capitalism*.[7]
- Testing is emphasized as a basis for validation and selection during the Conceptual Design, Preliminary Design and Detail Design phases. Testing involves a variety of mathematical, computer and physical modelling, plus monitoring, to ensure that the system fulfils the specifications and to rank the set of potential systems. For large or complex systems, final testing and optimization of the delivered system may be extended into the operation stage of the life-cycle.

Enhancing systems engineering through a whole system approach to help achieve sustainable design

Figure 3.3 shows the general Whole System Approach to design, which incorporates an emphasis on sustainability, as based on Systems Engineering. Feedback between phases is not shown, but steps can feed back to previous steps. There are three key terms used in Figure 3.3 and the subsequent process description, which are defined as follows:

1. *Design*: Selecting and integrating subsystems and components based on (1) an educated initial estimate of the optimal options for the system and (2) basic-level testing that verifies that the system will work;
2. *Optimize*: Refining the system composition based on analysis and testing so that specifications are best met; and
3. *Test*: Measuring the system performance using tools such as mathematical, computer and physical modelling, and monitoring, and comparing to the specifications.

The following sections provide a description of the basic WSD for the sustainability process. This process is complementary to the *Whole System Integration Process* developed by Bill Reed and colleagues.[9] There will be cases where some steps of the process are not relevant. In these cases, experience and common sense will dictate modifications to the process. Where a step is repeated in multiple phases, it is discussed and justified only in the first instance.

Phase 1: Need definition

The aim of the Need Definition phase is to develop an understanding of the system, its purpose and the attributes that will make it sustainable. There are three steps of Need Definition to consider:

1. service specification;
2. operating conditions specification; and
3. genuine targets specification.

Service specification

The service specification defines the service requirements (What services must the system provide?),

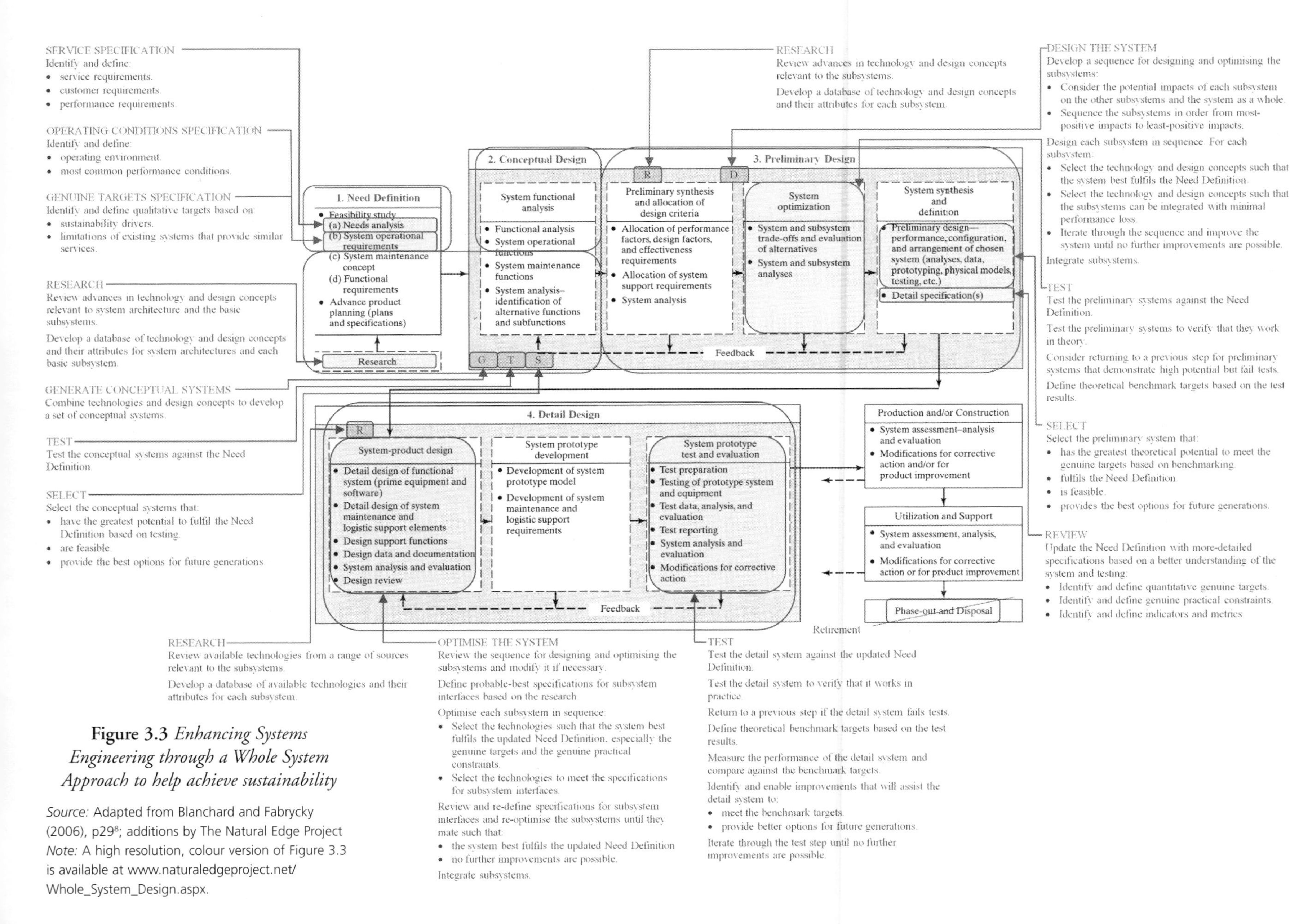

Figure 3.3 *Enhancing Systems Engineering through a Whole System Approach to help achieve sustainability*

Source: Adapted from Blanchard and Fabrycky (2006), p29[8]; additions by The Natural Edge Project
Note: A high resolution, colour version of Figure 3.3 is available at www.naturaledgeproject.net/Whole_System_Design.aspx.

customer requirements (What does the customer want the system to do?) and performance requirements (What are the measurable performance requirements?). Service requirements and customer requirements are used to derive performance requirements.

The customer requirements can be quite different to the service requirements. For example, while a service requirement for a modern car may be 'provides passenger mobility on roads up to 120km/h', a customer requirement may be 'provides passenger mobility on roads up to 220km/h'. In addition, not all customer requirements are available from the customer – analysis is sometimes required to ensure that the customer gets what they want, rather than what they specified. For example, some new water-efficient shower heads have been criticized by customers for not providing the sensation that 'enough' water is striking their heads. Had this requirement been identified earlier, the performance requirements may have defined a survey-based, user-satisfaction metric, in addition to the specified water flow rate and spread metrics.

Performance requirements relate to several system attributes, including mass, geometry, safety, reliability and ergonomics, to name just a few.

Operating conditions specification

The operating conditions specification defines the operating environment and most common performance conditions. An understanding of the operating environment is important, because the system does not act in isolation – it, like its subsystems, must be designed and optimized for the larger system of which it is a part. The operating environment conditions usually include temperature, pressure, humidity, geographic location, and the technological awareness and skill level of the local population. For example, a building designed for hot, dry central Australia is not suitable for temperate northern Canada. Another example: a typical pair of scissors with exposed blades is not suitable in a preschool.

Identifying the most common performance conditions will assist in right-sizing subsystems and thus in optimizing operating impacts. Most systems are designed and optimized for the maximum performance requirements. A better strategy is to design for the maximum performance requirements in order to ensure that the system can competently fulfil its services, and optimize for the most common performance requirements in order to optimize the operating impacts. For example, electrical equipment such as electric motors and power supplies have energy efficiency of 85–98 per cent near full load conditions and disproportionately reduced energy efficiencies at part load conditions. The operating energy costs of these units can easily eclipse the capital costs of the units themselves within months of continual use. Typically, electrical units are designed and optimized for full load plus extra load factored in as a precaution, but the system of which they are a part will most commonly operate at part load and hence very low efficiency. A better system may incorporate two units – a smaller electrical unit designed and optimized for the most common operating load and a secondary unit that only operates when required. The two smaller units will cost about the same as the single large unit but will operate near maximum efficiency and thus cost far less to operate.

Genuine targets specification

The genuine targets specification defines targets based on sustainability drivers and limitations of existing systems that provide similar services. The genuine targets specification defines targets based on both sustainability drivers and the limitations of existing systems that provide similar services. At the Need Definition phase, the targets are usually defined qualitatively because there is insufficient information to confidently define realistic, non-limiting quantitative targets, and in order to avoid defining arbitrary targets and constraints.

Sustainability is emphasized in the design process by defining specific targets from the Need Definition phase. Depending on the preferred definition of sustainability, the targets can, for example, be qualified as:

- minimize materials unrecovered;
- maximize energy efficiency:
- maximize bio-restoration; or
- minimize land use.

Additional targets are derived from the limitations of existing systems that provide similar services. That is, existing systems can be redesigned without their original limitations by simply defining targets that do not promote those limitations. For example, until the early 1990s, before standby energy consumption was a design issue for electrical appliances, fax machines consumed about the same amount of energy whether they were servicing a call or idling. The main issue was that fax machines spent about 90 per cent of the time idling, and so consumed about 90 per cent of their operating energy on non-service-providing tasks. By identifying this limitation, subsequent fax machines and electrical appliances were designed to consume relatively low amounts of energy when in idle mode. Limitations that are abundant in many modern systems include:

- excessive resource consumption;
- poor end-of-life options;
- built-in obsolescence;
- technological risks such as reliance on unstable resource sources; and
- social risks such as exposing humans and other organisms to toxic substances that can impair health.

Defining these targets emphasizes superior competitiveness over existing systems.

Phase 2: Conceptual design

The aim of the Conceptual Design phase is to thoroughly explore the solution space for all possible options that address the Need Definition, and to then generate a set of conceptual systems for further development. There are four steps of the Conceptual Design phase to consider:

1 research:
2 generate conceptual systems;
3 testing; and
4 selection.

Research

Research is critical in understanding the breadth of options available to address the Need Definition. In the Conceptual Design phase, the research focus is on reviewing advances in technology and design concepts relevant to system architecture and the basic subsystems, as new technologies that provide the same services can become outdated within six months. Design concepts include:

- Biomimetic design, which is nature- or bio-inspired system design;
- Green engineering and green chemistry, which are design for environmentally benign and restorative systems; and
- Design concepts that assist in developing novel and exciting systems.

System architectures can vary dramatically between systems that provide similar services. Basic subsystems refer to purposely broad and non-prescriptive subsystems, such as the 'power subsystem', 'structure subsystem', 'control subsystem' and 'communication subsystem'.

Brainstorming is used to encourage a creative design environment and assist in revealing all possible options – past, present and novel.

The review results are used to develop a database of technology and design concepts and their attributes for system architectures and each basic subsystem. Important attributes to research include resource requirements, operating environment, performance conditions, attributes identified in the performance requirements of the Need Definition, integratability and capital cost.

Generate conceptual systems

Technologies and design concepts are combined to generate a set of conceptual systems. A conceptual system can be graphically represented as a functional diagram, wherein subsystem blocks are connected by functional relationship links. External subsystems, such as resource pools and subsystems of the operating environment, are also represented as blocks. The links represent some relationship relevant to the Need Definition, such as 'impacts the mass of', 'impacts the ease of disassembly of' or 'impacts the concentration of'. Impacts include impacts through synergies and hidden impacts (see Chapter 4) and can be weighted with positive or negative magnitude using the information in the database.

Backcasting assists in developing conceptual systems that meet sustainability targets and provide options for future generations. Backcasting involves designing a 'future system' that is optimal in a sustainable future and then working backwards to develop a system that most closely matches the future system with currently available technologies. The outcome is that the developed system is the first model of a series on the path to a sustainable system.

Test

The set of conceptual systems are tested against the Need Definition. Tests reveal the approximate or at least relative potential for the conceptual systems to fulfil the Need Definition, including how far below or above the specifications the conceptual systems perform.

Select

The best or top few conceptual systems of the set are selected for further development and testing. Selection criteria include the potential to fulfil the Need Definition (based on testing), feasibility and the capacity to provide the best options for future generations.

Phase 3: Preliminary design

The aim of the Preliminary Design phase is to develop the set of conceptual systems into a set of preliminary systems by designing their subsystems such that the system as a whole best fulfils the Need Definition, and then to select the best system for further development. There are five steps of the Preliminary Design phase to consider for each conceptual system:

1. research;
2. designing the system;
3. testing;
4. selection; and
5. review.

Research

In the Preliminary Design phase, the research focus is on reviewing advances in technology and design concepts relevant to the subsystems. The subsystems refer to the types of technologies and design concepts that have been assigned to basic subsystems, such as the 'air pump', 'insulated wall', 'biomimetic mixing chamber', 'pressure sensor' and 'environmentally benign solvent'.

The review results are used to develop a database of technology and design concepts and their attributes for each subsystem.

Design the system

In designing the system, it is critical to design the subsystems in the right sequence. A flaw of the popular component engineering methodology, wherein subsystems are designed and optimized in isolation, is that committing to particular subsystem technologies without testing their impact on the system as a whole can result in an inherently non-optimizable system.

Thus a sequence for designing and optimizing the subsystems must be developed. The sequence relies on determining the potential impacts of each subsystem on the other subsystems and on the system as a whole. Updating the functional diagram with new information in the database assists in identifying the impacts.

The probable-best sequence will have subsystems in decreasing order of net positive impacts. That is, the sequence will start with the subsystem with the most positive impacts and end with the subsystem with the least positive impacts. Typically, the sequence correlates inversely with the transmission of resources through the system. That is, while the transmission of resources is from upstream (raw input) to downstream (end use), the sequence is usually downstream to upstream. However, the sequence is not necessarily linear; its general topology is a web.

Each subsystem is designed in sequence. The technology and design concepts for each subsystem are selected such that the system best fulfils the Need Definition. Usually, several combinations of technologies and design concepts can fulfil the Need Definition. In this case, emphasis is on meeting the genuine targets and providing options for future generations. Backcasting assists in designing a preliminary system that meets sustainability targets and provides options for future generations. In addition, technology and design concepts are selected such that the subsystems can be integrated with minimal performance loss.

Different technologies and design concepts have different impacts. Thus it is possible that designing, say, the last subsystem in the sequence will affect the suitability of the technology or design concept selected for the first subsystem. Consequently, multiple iterations through the sequence are usually required to ensure that no further improvements are possible and hence that the system best fulfils the Need Definition. The subsystems are then integrated to create a preliminary system.

Test

The preliminary system is tested against the Need Definition to verify that it works in theory. Emphasis is on testing the system as a whole. Tests reveal, with some accuracy and some degree of itemization, the potential for the preliminary system to fulfil the Need Definition. If the preliminary system fails any tests, there is value in considering revisiting previous steps of the design process to correct the faults. It is possible that some earlier decisions that seemed arbitrary at the time were, in fact, influential.

Tests are used to define theoretical benchmark targets. The targets reflect the theoretically optimal genuine targets. Benchmarking against theoretical testing is preferred over benchmarking against best practice, because best practice is rarely optimal.

Select

The best preliminary system is selected for further development and testing. Selection criteria include:

- the theoretical potential to meet the genuine targets (based on benchmarking);
- the capacity to fulfil the Need Definition;
- feasibility; and
- the capacity to provide the best options for future generations.

Review

Having now determined the subsystems technologies and design concepts, the Need Definition is reviewed and updated with more-detailed specifications. A better understanding of the system and testing assists to identify and define quantitative genuine targets, genuine practical constraints, and indicators and metrics.

Genuine targets can be upgraded from being defined qualitatively in terms of 'maximize' and 'minimize' to being defined quantitatively, such as 'achieve at least a factor five improvement in resource productivity', 'reduce greenhouse gas emission by 80 per cent' or 'reduce pollution by half'.

Genuine practical constraints are usually a consequence of customer requirements and the operating environment. Up until now, these constraints have only been passively considered, with only the relatively influential constraints being considered in the form of requirements. For the first time, the constraints are actively defined in detail for the purposes of providing a comprehensive Need Definition and hence undertaking meticulous assessment and optimization of the selected preliminary system. For example, where a preliminary system was required by the customer to 'replace the old blast furnace and fit in the same room', now a set of constraints can be defined as to the minimum distance between the system and human operators depending on the room geometry and the heat, ash emissions and gaseous emissions the system generates. Another example: where a preliminary cross-country cargo transport system was designed to 'avoid disturbing the local ecosystems', a set of constraints can now be defined as to the minimum distances between the system and certain ecosystems depending on the noise, vibration and air emissions that the system generates.

Indicators and metrics are quantitatively defined for every specification in the Need Definition. Typically, indicators and metrics are defined for internally measurable specifications, such as specifications for service output, reliability and operating life. It is important to also define indicators and metrics for externally measurable specifications, such as the environmental impact of input resources, the impact on the operating environment and the potential to provide options for future generations at end of life. The majority of sustainability driven specifications are externally measurable specifications. If indicators and metrics cannot be defined for a particular entry in the specification, then that entry is either vaguely defined, incorrectly defined or not relevant.

It is important to avoid defining targets, constraints, indicators and metrics arbitrarily.

Phase 4: Detail design

The aim of the Detail Design phase is to develop the selected preliminary system into the detail system by optimizing its subsystems such that the system as a whole best fulfils the updated Need Definition. There are three steps of the Detail Design phase to consider:

1 research;
2 optimizing the system; and
3 testing.

Research

In the Detail Design phase, the research focus is on reviewing available technologies from a range of sources relevant to the subsystems. The subsystems refer to the types of relatively specific technologies that have been assigned to subsystems, such as 'reverse cycle air-conditioner', 'proton exchange membrane fuel cell', 'wireless transceiver', 'sorting algorithm' or 'volatile organic compound-free fabric'. It is important to review a range of sources, because the performance and suitability of similar technologies can vary dramatically.

It is also important to obtain detailed specifications of the technologies, such as performance graphs, from sources. Basic performance specifications, such as nameplate specifications on electrical equipment, usually reveal very little information about performance at the most common operating condition or about the change in performance over time.

The review results are used to develop a database of technologies and their attributes for each subsystem. Important attributes to research include interfacing specifications and attributes derived from the updated Need Definition.

Optimize the system

In optimizing the system, it is critical to optimize the subsystems in the right sequence. Having now determined the subsystems technologies, the sequence for designing and optimizing the subsystems is reviewed and modified if necessary. Updating the functional diagram with new information in the database assists in identifying potential improvements to the sequence.

Equal emphasis is applied to optimizing both subsystems and the subsystem interfaces such that the system best fulfils the updated Need Definition. Direct interactions between subsystems are just as influential on the system performance as are the subsystems themselves. Both internal interfaces between subsystems and external interfaces to the operating environment are considered. Initially, the probable-best interface specifications are defined based on the information in the database.

Each subsystem is optimized in sequence. The technology for each subsystem is selected such that the system best fulfils the updated Need Definition. Any combination of technologies is now likely to meet the service specification and operating conditions specification. Thus emphasis is on meeting the now quantified genuine targets specification, meeting the new genuine practical constraints and providing options for future generations. In addition, technologies are selected to meet the interface specifications.

Since the initial interface specifications are not necessarily optimal, they must be reviewed and perhaps redefined based on their suitability to the subsystems technologies selected during the first iteration through the sequence. Different technologies and interface specifications have different impacts. Consequently, continuously redefining the interface specifications and multiple iterations through the sequence are usually required to ensure that no further improvements are possible and hence that the system best fulfils the updated Need Definition. The subsystems are then integrated to create the detail system.

Test

The detail system is tested against the updated Need Definition and tested to verify that it works in practice. Emphasis is on testing the system as a whole. Tests reveal, with good accuracy and itemization, the potential for the detail system to fulfil the Need Definition. If the detail system fails tests, previous steps of the design process must be revisited to correct the faults.

Tests are used to define theoretical benchmark targets. The targets reflect the theoretically optimal genuine targets and account for genuine practical constraints.

During testing, the performance of the detail system is measured and compared against the benchmark targets. It is unlikely that all attributes can be measured accurately using non-destructive processes. Some attributes are better estimated using alternative processes. Comparisons assist in meticulously identifying potential for improvements in the detail system towards meeting the benchmark targets and providing better options for future generations. The improvements are then incorporated into the system.

Iterating through the test step until no further improvements are possible will optimize the detail system for the available technology.

Elements of a Whole System Approach to Sustainable Design

There will be cases where some steps of the general Whole System Approach to design are not relevant. There are, however, ten key operational elements that leverage the greatest benefits. These elements are briefly discussed here, and further discussion about their practical use is given in Chapters 4 and 5. The ten key operational elements that are outlined below are integrated from lessons learnt by experienced designers, including Amory Lovins,[10] Hunter Lovins, Ernst von Weizsäcker, Bill McDonough, John Todd, Janis Birkeland[11] and Alan Pears[12] (see optional reading). These 10 key elements are widely seen as effective operational elements – many of them are consistent with operational principles and guidelines for engineering design developed by Dieter[13] and The International Council on Systems Engineering.[14]

Element 1: Ask the right questions

What is the required service? How can the service be provided optimally? Are there other possible approaches? For example, consumers want hygienic, comfortable containers out of which to drink. It makes no difference to consumers whether the container is made of glass or aluminium or if it is made of recycled material or not. It does, however, make a significant difference to the planet's ecosystems. This example demonstrates the potential impacts of decisions on the interpretation of the required service and the corresponding technologies and resource use. A service-based perspective assists in preventing arbitrary constraints, particularly on resource use.

Element 2: Benchmark against the optimal system

It is often useful to develop a simple functional model of the system, which assists the designer in thinking about the interacting components and to evaluate potential improvements to existing systems. The model is used to benchmark against both the theoretically and practically optimal systems. Benchmarking against 'best practice' is a dangerous strategy, as existing best practice is actually 'best of a bad lot of practice', because the reference cases typically were designed decades ago and the cost-effectiveness criteria for resource efficiency was likely to have been very stringent (less than a three-year payback period is a typical threshold). Today, it is possible to do much better.

Element 3: Design and optimize the whole system

In order to develop a system that meets the Need Definition with optimal resource use (energy, material and water inputs and outputs) and biological impact, it is important to consider all subsystems and their synergies, rather than single subsystems in isolation:

> Optimizing an entire system takes ingenuity, intuition and close attention to the way technical systems really work. It requires a sense of what's on the other side of the cost barrier and how to get to it by selectively relaxing your constraints. ... Whole-system engineering is back-to-the-drawing-board engineering. ... One of the great myths of our time is that technology has reached such an exalted plateau that only modest, incremental improvements remain to be made. The builders of steam locomotives and linotype machines probably felt the same way about their handiwork. The fact is, the more complex the technology, the richer the opportunities for improvement. There are huge systematic inefficiencies in our technologies; minimize them and you can reap huge dividends for your pocketbook and for the Earth. Why settle for small savings when you can tunnel through to big ones? Think big! (Rocky Mountain Institute, 1997)[15]

Element 4: Account for all measurable impacts

Modifications at the subsystem level can influence behaviour at the system level and achieve multiple benefits for single expenditures:

> This might seem obvious, but the trick is properly counting all the benefits. It's easy to get fixated on optimizing for energy savings, for example, and fail to take into account reduced capital costs, maintenance, risk or other attributes (such as mass, which in the case of a car, for instance, may make it possible for other components to be smaller, cheaper, lighter and so on). Another way to capture multiple benefits is to coordinate a retrofit with renovations that need to be done for other reasons anyway. Being alert to these possibilities requires lateral thinking and an awareness of how the whole system works. (Rocky Mountain Institute, 1997)[16]

Element 5: Design and optimize subsystems in the right sequence

Large improvements in resource use are, in many cases, a process of multiplying small savings in the right sequence. There is an optimal sequence for designing and optimizing the components of a system. The steps that yield the greatest impacts on the whole system should be performed first. For example, consider solar power for home energy supply. Solar cells are costly and provide perhaps one half or one third of the electricity consumed by a big heat pump striving to maintain indoor comfort in a building with an inefficient envelope, as well as inefficient glazing, lights and appliances. Suppose that the building was instead first made thermally insulated so it didn't need as big a heat pump (there are several much simpler ways to handle summer humidity). Then suppose the lights and appliances were made extremely efficient, with the latest technologies to reduce the house's total electric load. Now, the home's heating and cooling needs would be very small, its electrical needs could be met by only a few square metres of solar cells, and it would all work better and cost less.

Element 6: Design and optimize subsystems to achieve compounding resource savings

Life-cycle analysis shows that end-use resource efficiency is the most cost-effective way to achieve large improvements in resource use, because less resource demand at the end use creates opportunities to reduce resource demand throughout the whole supply chain:

> An engineer looks at an industrial pipe system and sees a series of compounding energy losses: the motor that drives the pump wastes a certain amount of electricity converting it to torque, the pump and coupling have their own inefficiencies, and the pipe, valves and fittings all have inherent frictions. So the engineer sizes the motor to overcome all these losses and deliver the required flow. But starting downstream – at the pipe instead of the pump – turns these losses into compounding savings. Make the pipe more efficient, and you reduce the cumulative energy requirements of every step upstream. You can then work back upstream, making each part smaller, simpler and cheaper, saving not only energy but also capital costs. And every unit of friction saved in the pipe saves about nine units of fuel and pollution at the power station. (Rocky Mountain Institute, 1997)[17]

Element 7: Review the system for potential improvements

It is important to identify potential resource use improvements and eliminate true waste (unrecovered resources) in each subsystem and at each stage of the life-cycle. For example, at most sites (from homes to large industrial plants) there is very limited measurement and monitoring of resource consumption at the process level, and rarely are there properly specified benchmarks. Thus plant operators rarely know where the potential for resource efficiency improvements lie. Monitoring and measurement help identify the system's performance inadequacies, which, when improved, can substantially improve performance and reduce costs. The integrated nature of systems provides a platform for the impact of individual improvements to compound and thus to generate an overall improvement greater than the sum of the individual impacts.

Element 8: Model the system

Mathematical, computer and physical models are valuable for addressing relatively complex engineering systems. For example, the Commonwealth Scientific and Industrial Research Organisation (CSIRO) has used computer modelling to make significant breakthroughs in fluid dynamics. Modelling of fluid dynamics by CSIRO is presenting opportunities for substantial efficiency improvements. A better understanding of liquid and gas flow has also helped CSIRO designers to improve the efficiency and performance of processing technologies in a wide range of applications. From such modelling, CSIRO has developed the Rotated Arc Mixer (RAM), which consumes five times less energy than conventional industrial mixers. The RAM is able to mix a range of fluids that were previously not mixable by other technologies.

Element 9: Track technology innovation

A key reason that there are still significant resource use improvements available through a Whole System Approach is that the rate of innovation in basic sciences and technologies has increased dramatically in the last few decades. Innovations in materials science in such things as insulation, lighting, super-windows, ultra-light metals and distributed energy options are creating new ways to re-optimize the design of old technologies. Innovation is so rapid that six months is now a long time in the world of technology. For example, consider the average refrigerator, for which most of the energy losses relate to heat transfer. The latest innovations in materials science in Europe have created a new insulation material that will allow refrigerators to consume 50 per cent less energy. Other examples include innovations in composite fibres that make it possible to design substantially lighter cars and innovations in light metals, which can now be used in all forms of transportation, from aircraft to trains to cars, and allow resource efficiency improvements throughout the whole system.

Element 10: Design to create future options[18]

A basic tenet of sustainability is that future generations should have the same level of life quality, environmental amenities and range of options as 'developed' societies enjoy today. It is also important to consider going beyond best practice and help create more options for future generations. It is crucial for designers to be aware of how new systems affect the options of future generations. For example, China is currently developing new coal-fired power stations at a rate of at least one per week. It is vital that these new power stations are correctly sited and designed to provide options for geo-sequestration of CO_2 emissions when the technology becomes commercially available.

There is now a wealth of literature on ways to achieve more sustainable designs through a Whole System Approach. Some of this literature is given in the optional reading section.

Optional reading

Birkeland, J. (2005) 'Design for ecosystem services: A new paradigm for ecodesign', presentation to SB05 Tokyo 'Action for Sustainability: The World Sustainable Building Conference', September

Blanchard, B. S. and Fabrycky, W. J. (2006) *Systems Engineering and Analysis* (fourth edition), Pearson Prentice Hall, Upper Saddle River, NJ, pp1–150

Dieter, G. E. (2000) *Engineering Design: A Materials and Processing Approach* (third edition), McGraw-Hill, Singapore, p228.

Hawken, P., Lovins, A. B. and Lovins, L. H. (1999) *Natural Capitalism: Creating the Next Industrial Revolution*, Earthscan, London, www.natcap.org, accessed 13 August 2007

International Council on Systems Engineering (1994) *A Process Description for a New Paradigm in Systems Engineering*, International Council on Systems Engineering, www.incose.org/ProductsPubs/pdf/techdata/PITC/ProcDescForNewParadigmForSE_1995-0810_SEPWG.pdf, accessed 11 July 2007

Lyle, J. (1999) *Design for Human Ecosystems*, Island Press, Washington, DC

Pears, A. (2004) 'Energy efficiency – Its potential: Some perspectives and experiences', background paper for International Energy Agency Energy Efficiency Workshop, Paris, April

Pears, A. (2005) 'Design for energy efficiency', presentation to Young Engineers Tasmania

Rocky Mountain Institute (1997) 'Cover story: Tunnelling through the cost barrier', *RMI Newsletter*, summer, pp1–4, www.rmi.org/images/other/Newsletter/NLRMIsum97.pdf, accessed 5 January 2007

Ulrich, K. T. and Eppinger, S. D. (2006) *Product Design and Development* (third edition), McGraw Hill

Van der Ryn, S. and Calthorpe, P. (1986) *Sustainable Communities: A New Design Synthesis for Cities, Suburbs and Towns*, Sierra Club Books, San Francisco, CA

Notes

1. Blanchard, B. S. and Fabrycky, W. J. (2006) *Systems Engineering and Analysis* (fourth edition), Pearson Prentice Hall, Upper Saddle River, NJ, p29. This is a textbook used in university undergraduate courses.
2. Blanchard, B. S. and Fabrycky, W. J. (2006) *Systems Engineering and Analysis* (fourth edition), Pearson Prentice Hall, Upper Saddle River, NJ, p39.
3. Blanchard, B. S. and Fabrycky, W. J. (2006) *Systems Engineering and Analysis* (fourth edition), Pearson Prentice Hall, Upper Saddle River, NJ, pp18–19.
4. Ulrich, K. T. and Eppinger, S. D. (2006) *Product Design and Development* (third edition), McGraw Hill.
5. Dieter, G. E. (2000) *Engineering Design: A Materials and Processing Approach* (third edition), McGraw-Hill, Singapore.
6. International Council on Systems Engineering (1994) *A Process Description for a New Paradigm in Systems Engineering*, International Council on Systems Engineering, www.incose.org/ProductsPubs/pdf/techdata/PITC/ProcDescForNewParadigmForSE_1995-0810_SEPWG.pdf, accessed 11 July 2007.
7. Hawken, P., Lovins, A. B. and Lovins, L. H. (1999) *Natural Capitalism: Creating the Next Industrial Revolution*, Earthscan, London, Chapter 6: 'Tunnelling through the cost barrier', www.natcap.org/images/other/NCchapter6.pdf, accessed 5 January 2007.
8. Blanchard, B. S. and Fabrycky, W. J. (2006) *Systems Engineering and Analysis* (fourth edition), Pearson Prentice Hall, Upper Saddle River, NJ, p29.
9. Reed, B., Boecker, J., Taylor, T., Pierce, D., Maine, G., Loker, R., Kessler, H., Borthwick, G., Culman, S.,

Batshalom, B., Settlemyre, K., Freehling, J., Sheffer, M., Martin, M., Toevs, B., Zurick, J., Mozina, T., Keiter, T., Albrecht, J., Montgomery, J., Prohov, R., Wardle, K., Dimond, D., Italiano, M., Gruder, S., Wong, M., Vujovic, V. and Swann, M. (2006) *Whole System Integration Process (WSIP) – Market Transformation to Sustainability Guideline Standard*, Report on workshop in Chicago, September 2006, www.integrativedesign.net/resources, accessed 28 September 2008.

10 RMI (1997) 'Tunnelling through the cost barrier', *RMI Newsletter*, summer, www.rmi.org/images/other/Newsletter/NLRMIsum97.pdf, accessed 5 January 2007.

11 Birkeland, J. (2005) 'Design for ecosystem services: A new paradigm for ecodesign', presentation to SB05 Tokyo 'Action for Sustainability: The World Sustainable Building Conference', September.

12 Pears, A. (2005) 'Design for energy efficiency', presentation to Young Engineers Tasmania; Pears, A. (2004) 'Energy efficiency – Its potential: Some perspectives and experiences', background paper for International Energy Agency Energy Efficiency Workshop, Paris.

13 Dieter, G. E. (2000) *Engineering Design: A Materials and Processing Approach* (third edition), McGraw-Hill, Singapore, p228.

14 International Council on Systems Engineering (1993) 'An identification of pragmatic principles: Final report', International Council on Systems Engineering, www.incose.org/ProductsPubs/pdf/techdata/PITC/PrinciplesPragmaticDefoe_1993-0123_PrinWG.pdf, accessed 11 July 2007.

15 RMI (1997) 'Tunnelling through the cost barrier', *RMI Newsletter*, summer, p3, www.rmi.org/images/other/Newsletter/NLRMIsum97.pdf, accessed 5 January 2007.

16 RMI (1997) 'Tunnelling through the cost barrier', *RMI Newsletter*, summer, p2, www.rmi.org/images/other/Newsletter/NLRMIsum97.pdf, accessed 5 January 2007.

17 RMI (1997) 'Tunnelling through the cost barrier', *RMI Newsletter*, summer, p3, www.rmi.org/images/other/Newsletter/NLRMIsum97.pdf, accessed 5 January 2007.

18 Birkeland, J. (2005) 'Design for ecosystem services: A new paradigm for ecodesign', presentation to SB05 Tokyo 'Action for Sustainability: The World Sustainable Building Conference', September.

4

Elements of Applying a Whole System Design Approach (Elements 1–5)

Educational aim

Despite the increasing popularity of books that promote Whole System Design (WSD), like *Natural Capitalism*, only a tiny fraction of technical designers routinely apply a WSD approach. Most conventionally trained technical designers will require, as an introduction to WSD, an operational guide to how the elements of a WSD approach can be applied. This Chapter presents a 'how-to' of the first five elements of WSD that were outlined in Chapter 3. The application of each element for optimal sustainability and competitive advantage is also discussed and then demonstrated with case studies.

Required reading

Adcock, R. (n.d.) 'Principles and practices of systems engineering', presentation, Cranfield University, UK, pp1–12, www.incose.org.uk/Downloads/AA01.1.4_Principles%20&%20practices%20of%20SE.pdf, accessed 2 July 2007

Hawken, P., Lovins, A. B. and Lovins, L. H. (1999) *Natural Capitalism: Creating the Next Industrial Revolution*, Earthscan, London, Chapter 6, pp111–124, www.natcap.org/images/other/NCchapter6.pdf, accessed 5 January 2007

Pears, A. (2004) 'Energy efficiency – Its potential: Some perspectives and experiences', background paper for International Energy Agency Energy Efficiency Workshop, Paris, pp1–16

Rocky Mountain Institute (1997) 'Tunnelling through the cost barrier', *RMI Newsletter*, summer, pp1–4, www.rmi.org/images/other/Newsletter/NLRMIsum97.pdf, accessed 5 January 2007

Introduction

The sustainability emphasis of taking a WSD approach brings environmental and social issues to the fore for consideration along with economic issues. The results to date of applying WSD consistently demonstrate environmental and social benefits, such as resource efficiency improvements by a factor of 2–10, pollution reduction, improved safety for users and the environment, and improved comfort for users compared to conventional systems. WSD also demonstrates economic benefits such as equal or reduced capital cost and reduced operating costs by a factor of 2–10 compared to conventional systems.

For these reasons, WSD is a powerful tool in achieving enhanced competitive advantage by reducing real costs and delivering quality systems. Despite increasing popularity, however, currently only a small fraction of technical designers routinely take a WSD approach. For many technical designers, a basic understanding of what a WSD approach is and what it can achieve can be attained by reviewing existing WSD case studies and literature. However, most case studies and literature do not accommodate an understanding of how to actually take a WSD approach in the design process.

It is challenging for many technical designers to make the unassisted mental leap from WSD case studies to a WSD approach. The main reason for this challenge is that conventional technical training typically focuses on specialized skills that are relevant to only a few subsystems or components – there is little foundation in systems thinking or systems engineering. Consequently, most conventionally trained technical designers will require, as an introduction to WSD, a clear understanding of the elements of a WSD approach. Chapter 5 builds on from this Chapter by presenting a 'how-to' of the final five of the ten elements of WSD.

Element 1: Ask the right questions

Taking a disciplined approach to questioning the need and suitability of each system feature and each step in the development process can greatly assist in developing an optimal system in a short period of time. Asking the right questions at the right time is the primary strategy for acquiring a deep understanding of the system and eliminating costly late modifications to it. The question that drives the system development process is 'How can a system be developed to provide the required service?'. Developing a system to provide a particular service is a process of working backwards to define the required system behaviours, researching available technologies and design concepts, and then selecting and integrating technologies and design concepts based on the Need Definition, as in Figure 4.1 (which also shows that social, technological and resource considerations drive questions throughout the development process).

An example of working backwards from the required service is illustrated in Figure 4.2, based on

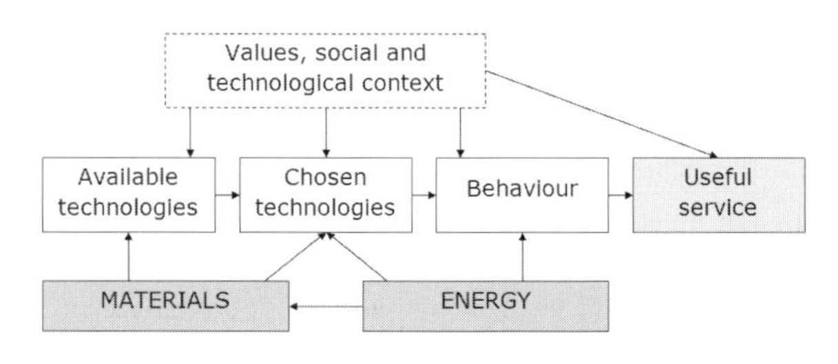

Source: Pears (2004)[1]

Figure 4.1 *A model of the resource and decisions inputs to providing a service*

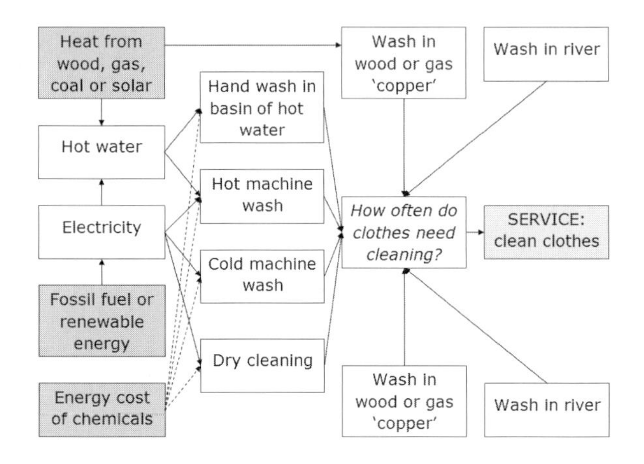

Source: Adapted from Pears (2004)[2]

Figure 4.2 *The range of potential technologies that can be used to provide the service of clean clothes, and the dependence of each technology on energy resources*

providing clean clothes. This figure shows the range of potential technologies that can emerge from a single question regarding the service of clean clothes, and the light grey boxes illustrate the unique dependence of each technology on energy resources.

There are three primary questions to ask during the Need Definition phase. These questions are critical for embedding a sustainability focus into the design specifications.

Question 1: What is the required service? (service specification)

A service is an intangible version of a good. Service provision engages goods and resources but does not necessarily consume any resource other than time. It is important to consider the number of times that the service is required during the system's operating life. Designing for too few services means the system could fail prematurely; designing for too many services means that more resources than necessary are engaged and lost.

Thinking in terms of service (pull) rather than product (push) eliminates the temptation to engage resources unnecessarily. For example, SafeChem,[3] a company established by The Dow Chemical Company, provides the service of cleaning metal components rather than providing solvents. SafeChem's solutions involve a closed-loop leasing system, wherein all solvents delivered to the customer are recovered, recycled and reused. Since any resources used to provide the service come at a cost, SafeChem have an incentive to develop the most resource-efficient solution. This service model strikes a stark contrast with the conventional product sales model, wherein the chemical supplier has an incentive to maximize the amount of solvent required. Calculations suggest that the enhanced model can reduce solvent consumption in already low-emission plants by 40 to 80 per cent.

Question 2: What is the optimal service? (genuine targets specification)

A single service can be provided by any number of systems. The role of a WSD approach is to ensure the service is provided optimally. The optimal service delivers the most benefits for the least cost while providing the required service. For the purposes of this chapter, we will be measuring the benefits and costs against environmental and economic sustainability. Social sustainability is another important consideration to measure against, but it is not discussed here, because its requirements vary substantially between systems, applications, regions and cultures.

Throughout its life, a theoretically optimal system will:

- Manage resources as in Table 4.1;
- Optimize the number of times its services are fulfilled;
- Be cost-effective; and
- Provide social fulfilment.

A system's 'life' extends from the development stage up to and including the system's final processing, after which the resources are no longer part of the system. This interpretation encompasses system upgrades and refurbishing.

Table 4.1 *Resource management for an optimal system*

Resource	Minimize	Maximize
Materials	Materials un-recovered Materials adversely disturbed	Materials upgraded Materials favourably dispersed
Energy	Energy un-recovered	Energy upgraded
Space	Space required	
Biological Impact	Toxic impact	Restorative impact

An example of a system that closely reflects Table 4.1 is Carnegie's Climatex®, a series of upholstery fabrics and its manufacturing process, developed in conjunction with MBDC. Climatex®[4] fabrics are coloured, biodegradable (*materials favourably dispersed, restorative impact*), non-toxic (*toxic impact*) and do not require additional chemical treatment. The fabrics contain pure wool from free-range sheep, polyester and organically grown ramie, and offer a combination of natural heat conservation, moisture absorption, good humidity transport and high elasticity. In developing the manufacturing process, approximately 1600 dye chemicals were tested, of which only 16 met ecological safety targets (*toxic impact*). Testing revealed that the outgoing process water, which is safe to drink, was cleaner than the incoming water (*materials upgraded, restorative impact*).[5] The biodegradable and non-toxic process material scrap is used to feed strawberry farms (*materials un-recovered*, restorative impact).[6]

Question 3: What are the system's operating conditions? (operating conditions specification)

There are two key considerations of a system's operational life:

1. A system always operates within a larger system, so it is important to account for external interactions.
2. Most systems are required to operate at several different loads and in several different environments, so it is important to optimize the system for the most common operating conditions while still designing the system to be reliable at maximum load.

There is an abundance of examples of these considerations, particularly the second, being overlooked. An example is air-conditioners: to ensure reliability, many air-conditioners are designed and optimized to operate at full load on extremely hot days. However, the most common operating conditions for air-conditioners are warm days at part load.

Another example is gas hot-water systems. Manufacturers design their gas hot-water systems in line with the Australian energy-rating scheme, which assumes that the average Australian home has 3–4 people and that hot water consumption is 200 litres per day. However, this profile is only relevant to

15 per cent of homes. Single-person homes make up 25 per cent of homes, while two-person homes make up an additional 25 per cent. Consequently, at least half of all gas hot-water systems provide twice the required capacity. The primary significance of this over-sizing is that standby heat losses are large in systems that maintain a full tank of hot water and increase with tank capacity. There is not yet a hot-water system small enough to match the needs of the 1–2-person home. This gap in the market presents a significant business opportunity for an entrepreneur.

Element 2: Benchmark against the optimal system

Benchmark targets are embedded in the system specifications. They are regularly consulted throughout the system development process, particularly to address questions and to evaluate the system during testing. It is important to *benchmark against the optimal system*, not merely against the best existing system. Most existing systems are sub-optimal and thus benchmarking against them can introduce arbitrary constraints that restrict the solution space. In addition, setting ambitious targets encourages breaking away from possibly restrictive cultural norms in order to explore new opportunities for service provision and system development.

An example of not benchmarking against the optimal system is early fax machines. In the early 1990s, it became apparent that fax machines were consuming as much energy in idle mode (which is 90 per cent of the time), as they did in servicing phone calls. Similar observations about other electrical and electronic appliances then began to emerge. These events led to a deliberate effort to improve standby operation by setting benchmark targets for low energy consumption. Today, there is an industry-wide benchmark of 1 Watt standby energy consumption in electrical and electronic appliances. This has led to significant cost savings and reductions in greenhouse gas emissions for businesses, governments, organizations and households.

An example of the opportunities for improvement through benchmarking is refrigerated supermarket display cabinets. The service of a typical open-case display cabinet is to cool and preserve meat and dairy products for health and safety reasons. Benchmarking a conventional cabinet against an optimal cabinet, which is developed using theoretical modelling and practical testing, reveals several performance differences and correspondingly several opportunities for improving the conventional cabinet. Firstly, the addition of a glass door over the cabinet's open face, similar to that of a supermarket freezer, can reduce energy consumption by 68 per cent, preserve food more effectively and deliver a host of other benefits.[7] The largest portion of this reduced energy consumption is from preventing the extreme leakage of cooling air from the open face. In addition, the otherwise leaked air contributes enormously to the supermarket's heating load and thus may unnecessarily increase the required capacity of the space heating system. The open face allows moist surrounding air to flow into the case and hence create a very high demand for defrost energy of 0.9 kWh/litre. The additional moisture also complicates both internal and external temperature control. Secondly, these display cabinets are usually lit internally by inefficient lamps, which radiate heat directly above food items. Benchmarking shows that this heat not only contributes to the cooling load, but also creates abnormal thermal profiles around food products that are difficult to control and thus could lead to poorly preserved products. Finally, some cabinets incorporate inefficient fans and motors that contribute further to the cooling load.

A two-stage process for developing benchmark targets

In WSD, benchmark targets are generally developed in a two-stage process (although minor updates are usually made throughout the process as the understanding of the system improves):

1 Initial benchmark targets are developed during the Need Definition phase. At this stage, they reflect *theoretical optimal* service provision as determined by Element 1: Ask the right questions, and are largely qualitative.
2 The benchmark targets are reviewed at the end of the Preliminary Design phase. At this stage, *practical constraints* are superimposed onto the theoretical targets to determine the practical targets, many of which are quantitative. Ensuring that any practical constraints are genuine prevents restricting the solution space.

Case study: Brick making

A good example to demonstrate the difference between theoretical and practical benchmark targets is brick manufacturing. Brick manufacturing is a multistage process (see Figure 4.3) in which water is added during forming and removed primarily during drying and less so during firing. This example shows the theoretical and practical considerations for optimizing energy consumption with respect to drying a batch of 1000 stiff-mud bricks. Note that designing and optimizing a subsystem in isolation from the rest of the system, as this case study suggests, is counter to a WSD approach. In fact, isolating the drying process as the sole means of removing moisture from brick will most likely result in a suboptimal solution, because there is an opportunity to reduce the quantity of input water during the earlier forming process and another opportunity to remove moisture during the subsequent firing process.

The *theoretical benchmark targets* for removing moisture depend on the type of processes used for each step of the production phase. For example, drying is an evaporation process that requires a certain quantity of thermal energy to evaporate a given quantity of water. Firing, on the other hand, is a chemical process that requires a certain quantity of thermal energy at a given temperature to break and reform chemical bonds. The *theoretical target* for energy consumption in the drying process can be determined as follows:

- Evaporation of water requires 2.26 MJ/kg of latent heat of evaporation, plus about 100 MJ of sensible heat per 1000 bricks to increase the water temperature from about 20°C to 100°C.
- The quantity of moisture in bricks before drying depends on the forming method. Stiff-mud bricks are 12–15 per cent moisture by mass.
- Given that a batch of 1000 stiff-mud bricks has a mass of 3000 kg, the 12 per cent moisture would have a mass of 360 kg.
- Thus the theoretical target for heat energy required to remove all the water from 1000 stiff-mud bricks by drying is 2.26 MJ/kg × 360 kg + 100 MJ = 914 MJ.

There are a few *practical constraints* in the drying process:

- The rate of removing moisture depends on the circulation and humidity of air around the bricks, while circulation of air depends on the spacing of the bricks and the size and spacing of the holes in the bricks; thus there are spacing and hole size constraints.
- The consistency of the bricks may not be uniform, so drying time may be extended to allow insulated regions of moisture to evaporate.
- The kiln and cart absorb some heat energy at a rate dependent on their heat capacity and temperature, so additional heat energy is required for cold starts.
- The consistency of bricks located near the cart may be affected by the different local thermal profile.

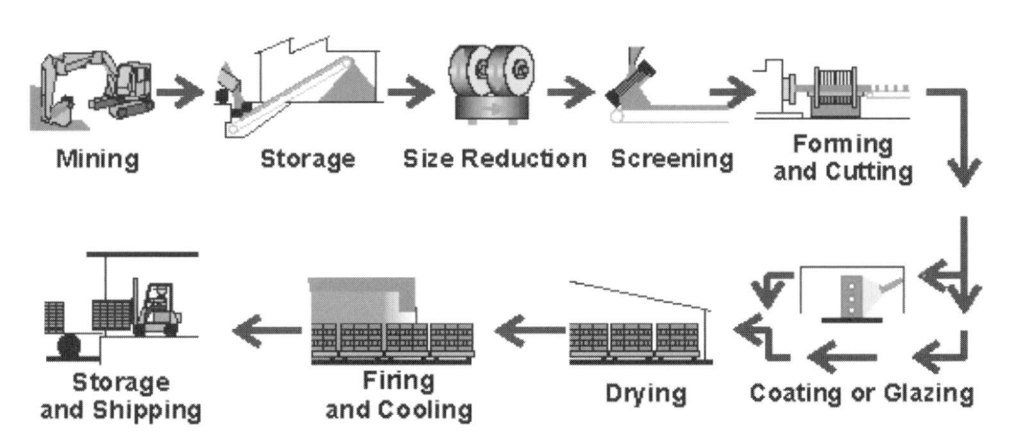

Source: The Brick Industry Association (2006)[8]

Figure 4.3 *The brick manufacturing process*

Superimposing these constraints onto the theoretical targets gives the *practical targets*. Note that these constraints may only be genuine constraints if their inclusion in the system was established in earlier phases of the development process. Constraints related to the cart affecting brick consistency, for example, is only genuine if the system must incorporate a cart.

Element 3: Design and optimize the whole system

Many systems and technologies are believed to be so complex and refined that only incremental improvements are possible. However, a WSD approach reveals that the more complex a technology is, the more opportunities there are for improvements.

Cleansheet design

System development is facilitated by *cleansheet design*. Cleansheet design is the process of developing a system from only a set of requirements and a 'clean sheet of paper'. Cleansheet design creates a design environment that offers the flexibility and creative space to innovate by investigating options that lie outside the bounds of the typical systems. In this environment, the system can be designed and optimized as a whole, and consequently *compromise can be minimized*. Cleansheet design *discourages defining arbitrary constraints*, which can promote compromise and premature commitment to a particular solution, and hence lead to an inherently non-optimizable system. A special case of setting arbitrary constraints is developing a system based on a previous version.

Designing the system as a whole

Designing the system as a whole is of value primarily during the Conceptual Design and Preliminary Design phases. It involves taking a *system-level emphasis on selecting and integrating subsystems and hence allowing synergies between subsystems to be identified and optimized*. System-level design is assisted by multidisciplinary development teams that have expertise in a broad range of relevant technology and design concepts.

An example of substantial benefits through optimizing synergies between subsystems is the IDEX™ kiln.[9] The IDEX is an energy-efficient, indirect-fired, controlled-atmosphere kiln developed by Wabash Alloys, an aluminium recycler and provider of aluminium alloy, in conjunction with Energy Research Company (ERCo). It optimizes the synergies between subsystems to:

- Clean aluminium with lower energy consumption;
- Have lower volatile organic compound (VOC) emissions;
- Reduce aluminium loss; and
- Better process unwanted substances than comparable kilns.

System-level design provides the opportunities for reducing energy consumptions that may not be *identified* with subsystem-level design (component design). The primary opportunities are reusing the IDEX's own heat to remove organic contaminants, hence reducing the IDEX's energy consumption, as well as being able to use cleaned, preheated aluminium scrap from the IDEX in the kiln at the next step of the process, which reduces the kiln's heating energy. If air leaks are eliminated and preheated scrap is used, the IDEX could save the aluminium recycling market three trillion British Thermal Units annually. Using the IDEX's cleaned and preheated aluminium in the furnace can also reduce aluminium loss and other metal particulate emissions by up to 34 per cent. Emissions from the IDEX are 5–50 times lower than stipulated by the New York State Department of Environmental Conservation Standards.

Optimizing the system as a whole

Optimizing the system as a whole is of value primarily during the Detail Design phase. It involves *comparing subsystem modifications against changes in both system service-provision and subsystem functionality*, and is usually an iterative process. Comparing against the system and subsystem levels ensures that both the synergies and subsystems are optimized for the benefit of the system. In the case of a contradiction, optimizing system service-provision is more important than optimizing the subsystem functionality.

Case study: Passenger vehicle design

An example of applying Element 3's features to transport vehicles is the *Hypercar Revolution*. Hypercar used *cleansheet design* to develop the Revolution concept vehicle[10] – a safe, well-performing, ultra-efficient passenger vehicle that is almost fully recyclable and competitively priced compared to current popular cars. Hypercar, having removed the *constraints* associated with upgrading past car models, was also able to overcome the *compromises* of conventional automobile design, such as making cars 'light or safe' and 'efficient or spacious'. Hypercar took advantage of *synergies between subsystems* to make the Revolution 'light *and* safe' and 'efficient *and* spacious'.

Hypercar identified a *synergy* between the mass of the Revolution's primary structure and almost every other subsystem, as in Figure 4.4 (a). Hypercar determined that a low mass primary structure would ultimately help meet ambitious targets for cost, as in Figure 4.4 (b), safety, efficiency, performance and comfort.[11]

The primary structure is made from an advanced composite material and consists of only 14 major components – about 65 per cent fewer than that of a conventional, stamped steel structure.[12] The components are made using a manufacturing process called Fibreforge™. Fibreforge requires very few sharp bends or deep draws, and thus has low tooling costs, high repeatability and fewer processing steps than conventional car assembly.[13] Unlike steel, the composite materials are lightweight, stiff, fatigue-resistant and rust-proof.[14] The higher cost of the composite materials is compensated for by a reduction in replacement parts and assembly complexity. The smoothly shaped shell components also reduce the Revolution's drag.

The low mass of the primary structure and low drag of the shell components reduce the mechanical load on the chassis and propulsion subsystems, which thus can be made relatively small and, again, low mass. Furthermore, some technological substitutions also become viable. For example, relatively small and very energy-efficient electric motors can power the wheels and thus eliminate the need for axles and differentials.[15]

The Revolution's low mass (about half that of similar-sized conventional vehicle) requires only 35 kW for cruising.[16] The low power requirement makes viable ambient pressure fuel-cell technology, which emits only pure water vapour and which would be too expensive at a large capacity. The fuel cell operates efficiently since it is mainly for cruising and thus can be optimized to operate over a very small power range. Additional power for acceleration is drawn from auxiliary 35 kW batteries. The batteries are recharged by a regenerative braking subsystem,[17] which is coupled with the electric motors. The braking system reduces energy losses to the surroundings.

The subsystem synergies in the Revolution are similar to those in all transport vehicles. Trucks, trains, ships and aircraft all have the potential to double or triple their fuel efficiency by 2025 (compared to conventional vehicles) if they incorporate the following:

- Low mass primary structures;
- Low drag shells;
- Small, high-efficiency propulsion systems and drive trains; and
- Electronic control.[18]

Some organizations are already taking the first steps – for example, Boeing has announced that its new passenger aircraft, which incorporates low-mass composite materials, will consume 20 per cent less fuel than other comparable aircraft. Wal-Mart has announced that its fleet of 6800 heavy trucks, which incorporate low drag shells, will double fuel efficiency by 2015, saving US$494 million by 2020.[19]

Element 4: Account for all measurable impacts

A defining feature of a system is the fact that making a single modification will create at least one other impact beyond that modification. This feature can be leveraged to assist in developing an optimal system in a short period of time. This leverage is created by *designing and optimizing for the most positive impacts*, which is of value primarily during the Conceptual Design, Preliminary Design and Detail Design phases. The aim is to optimize as much of the system as possible, not just a particular subsystem, with each decision.

There are two general types of impacts to consider in determining a decision's true value:

Impacts through synergies

Impacts through synergies occur internally within subsystems other than the subsystem that was acted on. These impacts can manifest themselves throughout the life of the system. Impacts through synergies are not

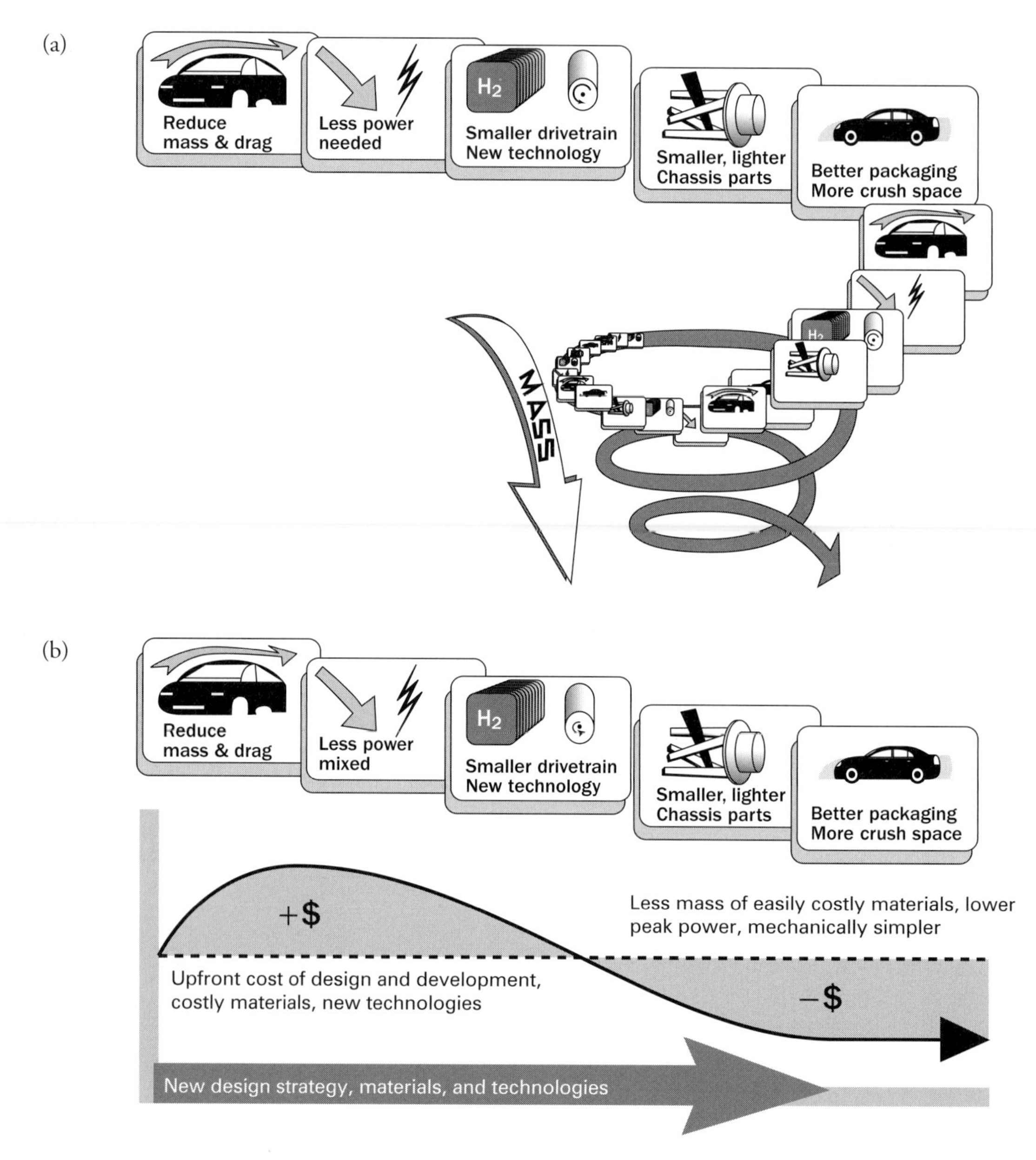

Source: Brylawski and Lovins (1998)[20]

Figure 4.4 *Potential (a) mass and (b) cost reductions through subsystem synergies arising from a low mass primary structure and low drag shell components in passenger vehicles*

always tangible, so further action on the subsystems of impact may be required to optimize the full impact of the original action.

An example that demonstrates accounting for *impacts through synergies* is RLX's blade server.[21] In 2001, RLX developed a blade server centred around an energy-efficient Transmeta processor. The Transmeta processor required 7–15W of electrical power while the leading competitor's equivalent processor required 75W. Using the Transmeta processor invoked several impacts through synergies, including the following:

- The high energy-efficiency of the Transmeta processor and other components led to RLX's server consuming 0.13A while the leading competitor's server consumed 79A to provide a similar service. The Transmeta processor generated relatively little heat and thus did not require auxiliary cooling through heat sinks and cooling fans (*energy-efficiency impacts on number of components*).
- Fewer cooling components and other component consolidation contributed to the size of the RLX server being 1/8 that of the leading competitor's server (*number of components impacts on volume*). Consequently, where RLX could house 336 of its servers in a rack, the leading competitor could house only 42, albeit with 2 processors per server (*volume impacts on number of racks*). Fewer components also contributed to the RLX server having a mass of 3lbs while the leading competitor's server had a mass of 29lbs (*number of components impacts on mass*).
- While the RLX server's capital cost may have been greater that of the leading competitor's server (depending on options), the total cost of ownership is about six times lower. The RLX server generated relatively little heat and thus required less cooling at the data centre level, which makes up about half of the total data centre operating cost (*energy efficiency impacts on cooling load*). The RLX server was also hot-pluggable with 12 times fewer Ethernet cables per rack, and did not require tools to install or remove, making management substantially easier (*number of components impacts on management complexity*).

Hidden impacts

Hidden impacts occur externally as a result of transforming resources into the state that they are used to create the system, including transportation. These impacts are relevant to input resources for both production and transportation. Hidden impacts can be measured using appropriate ecological and social indicators.

An example that demonstrates accounting for hidden *ecological* impacts is EnviroGLAS®.[22] EnviroGLAS create hard surfaces such as floors, countertops and landscaping materials from landfill-bound post-consumer and industrial ceramic materials. EnviroGLAS products are composed of about 75 per cent recycled glass or porcelain and 25 per cent epoxy binder by volume. EnviroGLAS's products have a favourable *hidden processing impact* compared to similar products made from virgin materials. The hidden impact of 1kg of recycled glass includes 550g of abiotic material and 7kg of water.[23] By contrast, the hidden impact of 1kg of virgin bottle glass includes 2.6kg of abiotic material and 13kg of water.[24] The higher consumption of hidden input resources for virgin bottle glass is attributed largely to two differences:

1. The disturbance of earth in order to obtain the silica sand, calcium, soda and magnesium ingredients; and
2. The high energy demand to heat the ingredients to 1500°C in a furnace.

EnviroGLAS's products also have a favourable *hidden transportation impact*. EnviroGLAS's materials are sourced from local recycling programmes, so fuel consumption, gas emissions and human labour costs are low. By contrast, each individual material input for virgin ceramics is transported separately over large distances, often internationally.

An example that demonstrates accounting for hidden *social* impacts is lead in electronic equipment. Most notably, lead makes up 40 per cent of electrical solder in printed circuit boards; it makes up 1–3kg of glass panes in cathode ray tube (CRT) monitors and televisions;[25] and it comprises a substantial portion of mass in hard drives. If not properly managed during production and at end-of-life, lead can be consumed by humans, other animals, plants and micro-organisms in a number of ways, including as airborne dust, ash or fumes, as well as ions in water bodies.[26] Associated with lead consumption are negative *hidden social impacts*. Lead can accumulate in humans and cause a variety of health issues, including damage to the nervous system, blood system, kidneys, and reproductive system and impairment of children's brain development and consequent intellectual impairment.[27] The risk of lead consumption is high since a large portion of waste electronic equipment is exported to developing countries for 'recycling', which often involves people sorting through the waste in landfills[28] or workshops[29] without protection. Lead-free alternatives for both electrical solder and CRT monitors and televisions are now available. Lead-free solder replaces lead with a number of other metals and is being used for many electronic products. In addition, liquid crystal display (LCD) monitors use plastic instead of glass panels, and many plasma televisions use lead-free glass.

Ecological indicators

Ecological indicators are tools that estimate the hidden ecological impacts of resources. Several quantitative and qualitative ecological indicators have been developed. With so many similar ecological indicators available, it is important to understand the underlying assumptions and ensure that the selected (or developed) indicators are effective for the application. Generally, effective ecological indicators meet the following conditions:[30]

- They are simple, yet reflect essential environmental stress factors;
- They are scientifically defensible, albeit not necessarily scientifically complete;
- They are based on characteristics that are common to all processes, goods and services;
- The selected characteristics are straightforwardly measurable or calculable, irrespective of geographic location;
- Obtaining results with these measures is cost-effective and timely;
- The measures permit the transparent and reproducible estimation of environmental stress potentials of all conceivable plans, processes, goods and services throughout the system's life;
- Their use always yields directionally safe answers;
- They form a bridge to economic models; and
- They are acceptable and usable at all levels: local, regional and global.

No single ecological indicator estimates all hidden ecological impacts. Hence, sets of ecological indicators are compiled to comprehensively estimate a decision's hidden impacts. Accurate estimations rely on sets of ecological indicators that do not overlap.

Case study: Post office design

An example of applying Element 4's features to commercial buildings is the Reno post office.[31] Reno post office is a modern warehouse with high ceilings and black floors. It houses two noisy mail sorting machines, which are so tedious and stressful to use that an operator can only work on a machine for 30 minutes at a time. In the early 1980s, the post office was renovated at a cost of US$300,000 with the aim of improving energy efficiency. The renovation consisted of two parts, both of which invoked impacts through synergies on other subsystems:

1 Lowering and sloping the ceiling; and
2 Installing energy-efficient, longer-lasting, softer-light lamps.

Lowering the ceiling reduced the volume of space that required heating and cooling and thus reduced the energy demand for heating and air-conditioning. Sloping the ceiling enhanced indirect lighting and reduced the need for direct lighting. The new ceiling also improved the warehouse's acoustic properties. The new lamps reduced direct energy consumption and replacement costs. They also emit less heat and therefore reduce the energy demand for space heating. In total, the energy cost savings from these two activities amounted to US$22,400 per year, while additional maintenance costs savings amounted to US$30,000 per year. The reduced demand for energy and paint reduced the hidden impacts of supplying those resource inputs, as estimated by the ecological indicators.

The renovations also invoked some unplanned and more valuable impacts on other subsystems of the post office. Productivity increased by 6 per cent and the frequency of mail sorting errors dropped to the lowest error rate in the region. The productivity improvement yielded savings of US$400,000–500,000 per year.

Element 5: Design and optimize subsystems in the right sequence

While a WSD approach emphasizes system-level design and optimization, there is a clear role for subsystem-level design and optimization, primarily in the

Preliminary Design and Detail Design phases. In these phases, the *management of subsystem-level design and optimization is determined by the system synergies.* Consistent with Element 4: Account for all measurable impacts, the decision to design or optimize a certain subsystem at a given time in the development process depends on its potential impact. Decisions that have the most-positive impact are favoured. An extension of this logic is that there is a single subsystem or small number of subsystems that are best designed or optimized first, and also that there is a logical sequence, or set of parallel sequences, for designing or optimizing subsystems that will yield the optimal system with minimal effort and human resource cost. The sequence is generally non-linear and usually iterative. Visual aids, such as schematic system maps, assist in identifying the sequence.

In practice, subsystem-level design and optimization is manifested as the following generally mutually exclusive sequences:[32]

- People before hardware;
- Shell before contents;
- Application before equipment;
- Quality before quantity;
- Passive before active; and
- Demand before supply.

Case study: Office building design

An example of applying Element 5's features to commercial buildings is *Green on the Grand*,[33] a two-storey office property in the US, developed in 1995. The subsystem with the *most positive potential impact* with respect to energy is the building's envelope (*shell*). Green on the Grand is orientated to the south and its form helps optimize (*passive*) daylighting and solar gain.

The next subsystem with positive potential impact with respect to energy is the building's lighting subsystem. Lighting (*application*) is primarily provided by (*passive*) daylight. Daylight penetration and the associated glare and heat gain are optimized using appropriately located large windows, glazed entranceways and interior glass walls. In summer, fabric roller-blinds and horizontal blinds with slats provide shading while deflecting light into offices. Secondary lighting requirements are met with (*active*) dimmable, photo- and motion-controlled energy-efficient lamps (*equipment*). The lamps are located to emphasize task lighting (*quality*). Lighting electricity consumption (*demand*) is 50 per cent lower than for conventional offices.

The next subsystem with positive potential impact with respect to energy is the building's climate control. Climate control (*application*) is primarily provided by the (*passive*) ventilation subsystem. The ventilation subsystem is mainly composed of two heat exchangers, two fans and a heating/cooling coil. The subsystem supplies offices with fresh outdoor air while displacing stale air. Secondary climate control requirements are met with the (*active*) radiant heating and cooling subsystem (*equipment*). Compared with forced air subsystems, radiant subsystems have lower motor energy consumption and fewer moving components, and are more energy efficient. For example, the subsystem includes water-based radiant panels that cover 30 per cent of the ceiling area. The panels warm the offices in winter, operating at 35°C, and cool the offices in summer, operating at 13°C.

The subsystem with the most-positive potential impact with respect to water is the building's water system. Water services are provided through a number of subsystems. The bathrooms are equipped with water-efficient fixtures, sensor-controlled shower heads and urinals (*quality*), and water-efficient toilets. The bathrooms are centrally located, which reduces hot water consumption (*demand*) by 20 per cent. Water is heated by a high-efficiency gas boiler. Rainwater is used for irrigation and in the cooling pond, which helps expel excess heat in summer. The water consumption (*demand*) of Green on the Grand is 30 per cent lower than for conventional offices.

Optional reading

Birkeland, J. (2002) *Design for Sustainability: A Sourcebook of Ecological Design Solutions*, Earthscan, London

Çengel, Y. and Boles, M. (2008) *Thermodynamics – An Engineering Approach* (6th edition), McGraw-Hill, pp78–96, http://highered.mcgraw-hill.com/classware/infoCenter.do?isbn=0073529214&navclick=true, accessed 28 August 2008

Hawken, P., Lovins, A. B. and Lovins, L. H. (1999) *Natural Capitalism: Creating the Next Industrial Revolution*, Earthscan, London, Chapter 6: 'Tunnelling through the cost barrier', www.natcap.org/images/other/NCchapter6.pdf, accessed 26 November 2006

Lovins, A. B., Datta, E. K., Bustnes, O. E., Koomey, J. G. and Glasgow, N. J. (2004) *Winning the Oil Endgame: Innovation for Profits, Jobs and Security, Technical Annex*, Rocky Mountain Institute, Snowmass, CO, www.oilendgame.com/TechAnnex.html, accessed 29 July 2007

McDonough, W. and Braungart, M. (2002) *Cradle to Cradle: Remaking the Way We Make Things*, North Point Press, New York

Pears, A. (2004) 'Energy efficiency – Its potential: Some perspectives and experiences', background paper for International Energy Agency Energy Efficiency Workshop, Paris, April, www.naturaledgeproject.net/Documents/IEAENEFFICbackgroundpaperPearsFinal.pdf, accessed 14 August 2007

Romm, J. J. and Browning, W. D. (1998) *Greening the Building and the Bottom Line*, Rocky Mountain Institute, CO, www.rmi.org/images/PDFs/BuildingsLand/D94-27_GBBL.pdf, accessed 30 July 2007

Van der Ryn Architects (n.d.) 'Five principles of ecological design', available at Van der Ryn Architects website, http://64.143.175.55/va/index-methods.html, accessed 14 August 2007

Van der Ryn, S. and Cowan, S. (1995) *Ecological Design*, Island Press, New York

Van der Ryn, S. (2005) *Design for Life: The Architecture of Sim Van der Ryn*, Gibbs-Smith Publishers, New York

Von Weizsäcker, E., Lovins, A. B. and Lovins, L. H. (1997) *Factor Four: Doubling Wealth, Halving Resource Use*, Earthscan, London

William McDonough Architects (1992) *Hanover Principles of Design for Sustainability*, prepared for EXPO 2000, The World's Fair, Hanover, Germany, www.mcdonough.com/principles.pdf, accessed 14 August 2007

Notes

1. Pears, A. (2004) 'Energy efficiency – Its potential: Some perspectives and experiences', background paper for International Energy Agency Energy Efficiency Workshop, Paris, p8, www.naturaledgeproject.net/Documents/IEAENEFFICbackgroundpaperPearsFinal.pdf, accessed 30 March 2008.
2. Pears, A. (2004) 'Energy efficiency – Its potential: Some perspectives and experiences', background paper for International Energy Agency Energy Efficiency Workshop, Paris, p9, www.naturaledgeproject.net/Documents/IEAENEFFICbackgroundpaperPearsFinal.pdf, accessed 30 March 2008.
3. SafeChem (2005) 'Chemical leasing within the SAFECHEM business model', SafeChem, www.dow.com/safechem/about/news/20050916.htm, accessed 30 November 2006.
4. See Climatex, 'Products', at www.climatex.com/en/products/products_overview.html, accessed 28 November 2006.
5. Pollock Shea, C. (n.d.) 'Mimicking nature and designing out waste', Worldwatch Institute, http://sustainability.unc.edu/index.asp?Type=Materials&Doc=wasteDesigningItOut, accessed 28 November 2006.
6. Steelcase (2006) 'Living in a material world', Steelcase, www.360steelcase.com/e_000150202000035897.cfm?x=b11,0,w, accessed 28 November 2006.
7. Faramarzi, R., Coburn, B. and Sarhadian, R. (2002) *Performance and Energy Impact of Installing Glass Doors on an Open Vertical Deli/Dairy Display Case*, ASHRAE.
8. The Brick Industry Association (2006) 'Manufacturing of brick', *Technical Notes on Brick Construction*, no 9, The Brick Industry Association, Reston, VA, p2, www.brickinfo.org/bia/technotes/t9.pdf, accessed 31 March 2008.
9. See US Department of Energy, 'Indirect-fired kiln conserves scrap aluminum and cuts costs', *Energy Matters*, at www1.eere.energy.gov/industry/bestpractices/pdfs/em_proheat_firedkiln.pdf, accessed 26 November 2006.
10. Lovins, A. B. and Cramer, D. R. (2004) 'Hypercars, hydrogen, and the automotive transition', *International Journal of Vehicle Design*, vol 35, nos 1–2, p54.
11. Brylawski, M. M. and Lovins, A. B. (1998) *Advanced Composites: The Car is at the Cross-Roads*, RMI, p6, www.rmi.org/images/other/Trans/T98-01_CarAtCrossroads.pdf, accessed 19 January 2006; Williams, B. D., Moore, T. C. and Lovins, A. B. (1997) *Speeding the Transition: Designing a Fuel-Cell Hypercar*, Rocky Mountain Institute, p2, www.rmi.org/images/other/Trans/T97-09_SpeedingTrans.pdf, accessed 14 January 2005.
12. Lovins, A. B. and Cramer, D. R. (2004) 'Hypercars, hydrogen, and the automotive transition', *International Journal of Vehicle Design*, vol 35, nos 1–2, p65.
13. Lovins, A. B. and Cramer, D. R. (2004) 'Hypercars, hydrogen, and the automotive transition', *International Journal of Vehicle Design*, vol 35, nos 1–2, p66.
14. Cramer, D. R. and Brylawski, M. M. (1996) *Ultralight-Hybrid Vehicle Design: Implications for the Recycling Industry*, Rocky Mountain Institute, p2, www.rmi.org/images/other/Trans/T96-14_UHVDRecycleInd.pdf, accessed 10 July 2005.
15. Moore, T. C. (1996) 'Ultralight hybrid vehicles: Principles and design', paper presented at 13th International Electric Vehicle Symposium, Osaka, Japan, www.rmi.org/images/other/Trans/T96-10_UHVPrinDsn.pdf, accessed 11 July 2005.

16 Moore, T. C. (1996) 'Ultralight hybrid vehicles: Principles and design', paper presented at 13th International Electric Vehicle Symposium, Osaka, Japan, www.rmi.org/images/other/Trans/T96-10_UHVPrinDsn.pdf, accessed 11 July 2005.
17 Lovins, A. B. and Cramer, D. R. (2004) 'Hypercars, hydrogen, and the automotive transition', *International Journal of Vehicle Design*, vol 35, nos 1–2, pp50–85.
18 Lovins, A. B., Datta, E. K., Bustnes, O. E., Koomey, J. G. and Glasgow, N. J. (2004) *Winning the Oil Endgame: Innovation for Profits, Jobs and Security, Technical Annex*, Rocky Mountain Institute, Snowmass, CO, www.oilendgame.com/TechAnnex.html, accessed 29 July 2007.
19 Rocky Mountain Institute (2007) 'Wal-Mart announces plans to double its heavy-duty truck fleet's fuel efficiency, RMI, www.rmi.org/store/p15details10.php?x=1&pagePath=00000000, accessed 29 July 2007.
20 Brylawski, M. M. and Lovins, A. B. (1998) *Advanced Composites: The Car is at the Cross-Roads*, RMI, pp5–6, www.rmi.org/images/other/Trans/T98-01_CarAtCrossroads.pdf, accessed 19 January 2006; Williams, B. D., Moore, T. C. and Lovins, A. B. (1997) *Speeding the Transition: Designing a Fuel-Cell Hypercar*, Rocky Mountain Institute, p3, www.rmi.org/images/other/Trans/T97-09_SpeedingTrans.pdf, accessed 14 January 2005.
21 Los Alamos National Laboratory (2001) *Supercomputing in Small Spaces*, Los Alamos National Laboratory, http://public.lanl.gov/radiant/pubs/sss/sc2001-pamphlet.pdf, accessed 9 January 2005; Transmeta Corporation (2002) 'Transmeta's 1GHz Crusoe Processor enables fast, high density blade server from RLX Technologies', Transmeta Corparation, http://investor.transmeta.com/ReleaseDetail.cfm?ReleaseID=97691, accessed 30 March 2007.
22 EnviroGLAS Products (2005) *Totally Cool!*, EnviroGLAS Products, www.enviroglasproducts.com/news-totallycool.html, accessed 30 November 2006.
23 Sorensen, J. and NOAH Sustainability Group (2005) *Ecological Rucksack for Materials Used in Everyday Products*, NOAH, www.noah.dk/baeredygtig/rucksack/rucksack.pdf, accessed 20 November 2006.
24 Sorensen, J. and NOAH Sustainability Group (2005) *Ecological Rucksack for Materials Used in Everyday Products*, NOAH, www.noah.dk/baeredygtig/rucksack/rucksack.pdf, accessed 20 November 2006.
25 OECD (2003), cited in Brigden, K., Labunska, I., Santillo, D. and Allsopp, M. (2005) *Recycling of Electronic Wastes in China and India: Workplace and Environmental Contamination*, Greenpeace International, p26, www.greenpeace.org/raw/content/india/press/reports/recycling-of-electronic-wastes.pdf, accessed 9 July 2006.
26 Brigden, K., Labunska, I., Santillo, D. and Allsopp, M. (2005) *Recycling of Electronic Wastes in China and India: Workplace and Environmental Contamination*, Greenpeace International, p26, www.greenpeace.org/raw/content/india/press/reports/recycling-of-electronic-wastes.pdf, accessed 9 July 2006.
27 Brigden, K., Labunska, I., Santillo, D. and Allsopp, M. (2005) *Recycling of Electronic Wastes in China and India: Workplace and Environmental Contamination*, Greenpeace International, p26, www.greenpeace.org/raw/content/india/press/reports/recycling-of-electronic-wastes.pdf, accessed 9 July 2006; Environment Victoria (2005) *Environmental Report Card on Computers 2005: Computer Waste in Australia and the Case for Producer Responsibility*, Environment Victoria, p8, www.envict.org.au/file/Ewaste_report_card.pdf, accessed 9 July 2006.
28 Puckett, J., Byster, L., Westervelt, S., Gutierrez, R., Davis, S., Hussain, A. and Dutta, M. (2002) *Exporting Harm: The High-Tech Trashing of Asia*, Basel Action Network, www.ban.org/E-waste/technotrashfinalcomp.pdf, accessed 18 September 2007; Puckett, J., Westervelt, S., Gutierrez, R. and Takamiya, Y. (2005) *The Digital Dump: Exporting Re-Use and Abuse to Africa*, Basel Action Network, www.ban.org/BANreports/10-24-05/documents/TheDigitalDump.pdf, accessed 18 September 2007.
29 Brigden, K., Labunska, I., Santillo, D. and Allsopp, M. (2005) *Recycling of Electronic Wastes in China and India: Workplace and Environmental Contamination*, Greenpeace International, p3, www.greenpeace.org/raw/content/india/press/reports/recycling-of-electronic-wastes.pdf, accessed 18 September 2007.
30 Schmidt-Bleek, F. B. (1999) *Factor 10: Making Sustainability Accountable, Putting Resource Productivity into Practice*, Factor10 Institute, p68, www.factor10-institute.org/pdf/F10REPORT.pdf, accessed 11 April 2007.
31 Romm, J. J. and Browning, W. D. (1998) *Greening the Building and the Bottom Line*, Rocky Mountain Institute, www.rmi.org/images/other/GDS/D94-27_GBBL.pdf, accessed 30 July 2007.
32 Rocky Mountain Institute (1997) 'Tunnelling through the cost barrier: Why big savings often cost less than small ones', *Rocky Mountain Institute Newsletter*, vol 13, no 2, p3, www.rmi.org/images/other/Newsletter/NLRMIsum97.pdf, accessed 6 June 2007.
33 Royal Institute of Chartered Surveyors (2005) *Green Value: Growing buildings, Growing Assets – Case Studies*, Royal Institute of Chartered Surveyors, pp7–10, http://rics.org/NR/rdonlyres/4CB60C80-C5E9-46F4-8D0A-D9D33B7A2594/0/GreenValueCaseStudies.pdf, accessed 6 June 2007.

5

Elements of Applying a Whole System Design Approach (Elements 6–10)

Educational aim

Chapter 5 builds on from Chapter 4 to describe the final five elements of applying a Whole System Design (WSD) approach. It presents a 'how-to' of the last five elements of WSD. The application of each element for optimal sustainability and competitive advantage is discussed and then demonstrated with case studies.

Required reading

Adcock, R. (n.d.) 'Principles and practices of systems engineering', presentation, Cranfield University, UK, pp1–12, www.incose.org.uk/Downloads/AA01.1.4_Principles%20&%20practices%20of%20SE.pdf, accessed 2 July 2007

Hawken, P., Lovins, A. B. and Lovins, L. H. (1999) *Natural Capitalism: Creating the Next Industrial Revolution*, Earthscan, London, Chapter 6: 'Tunnelling through the cost barrier', pp111–124, www.natcap.org/images/other/NCchapter6.pdf, accessed 2 July 2007

Pears, A. (2004) 'Energy efficiency – Its potential: Some perspectives and experiences', background paper for International Energy Agency Energy Efficiency Workshop, Paris, pp1–16

Rocky Mountain Institute (1997) 'Tunnelling through the cost barrier', *RMI Newsletter*, summer, pp1–4, www.rmi.org/images/other/Newsletter/NLRMIsum97.pdf, accessed 5 January 2007

Element 6: Design and optimize subsystems to achieve compounding resource savings

Many systems have subsystem synergies that resemble a distinct 'path' originating in a single or small number of subsystems. As discussed in Element 5: Design and optimize subsystems in the right sequence, these subsystems usually have the most-positive impact and thus are best designed and optimized first. An important observation is that the sequences resulting from applying Element 5 are generally counter to the actual resource transmissions. That is, the subsystem design and optimization sequences in the Preliminary Design and Detail Design phases are a set of integrated, general *downstream to upstream sequences*.

In systems with actual resource transmissions, a downstream to upstream sequence is equivalent to the sequence 'demand before supply' in Element 5. The downstream to upstream sequence also applies more abstractly to systems with unidirectional subsystem synergies that do not represent resource transmissions.

The impacts of subsystems in series compound rather than sum. Thus it is important to design and optimize subsystems such that the compounded impact is optimized. Compounding impacts can be leveraged to turn several small improvements at the subsystem level into a large positive impact on the system.

Focusing on end-use efficiency can create a cascade of savings all the way back to the power plant, dam, mine or forest. This is why an engineering focus on a WSD approach to re-optimize 'end-use' engineered systems such as motors, HVAC systems, buildings and cars can help business and nations reduce costs of infrastructure and environmental pressures significantly.

It is by focusing on these engineered systems which actually provide the services we need, close to the end-user, that big energy savings can be achieved back up the line. Much electricity has been used to create and run these end-user engineered systems. Hence any savings in the amount of energy needed to run these end-use engineered systems will produce a cascade of savings back to the electrical power plant. Consider the example of motors. Motors use about 60 per cent of the world's electricity.[1] Those used in pumping applications use about 20 per cent of the world's energy.[2] So if it is possible to reduce the amount of energy that a motor system needs to provide, this will create a cascade of savings all the way back to the power plant.

This effect of compounding savings from improving the efficiency of an industrial pumping system is seen in Figure 5.1, which shows the energy transmission and losses from raw material to the service of a pumped fluid in a typical pumping system.

In this case, the energy losses compound at every subsystem downstream of the 'Fuel energy input' until only 9.5 per cent of the original input energy remains to provide the service. However, designing and optimizing this system in the counter sequence, downstream to upstream, creates an opportunity to turn compounding energy losses (a negative impact) into compounding energy reductions (a positive impact). The subsystem furthest downstream is the end use. Reducing 'Energy output' by 1 unit eliminates 10.5 (= 100/9.5) units of 'Fuel energy input'. Next, reducing 'Pipe losses' by 1 unit eliminates a further 8.3 units; 'Throttle losses', 5.5 units; 'Pump losses', 4.2 units; and so on up to 'Power plant losses', which eliminates 1 unit of 'Fuel energy input'. The result is that reducing energy losses by 1 unit in each subsystem, or 8 units in total, eliminates 40 units of 'Fuel energy input'.

Case study: Solar cell design

Size and mass are two physical characteristics that limit the practicality of current photovoltaic (PV) systems. These characteristics are important because PV systems are often mounted on roofs or eaves, where space and structural integrity can be limited. Solar cells, the main

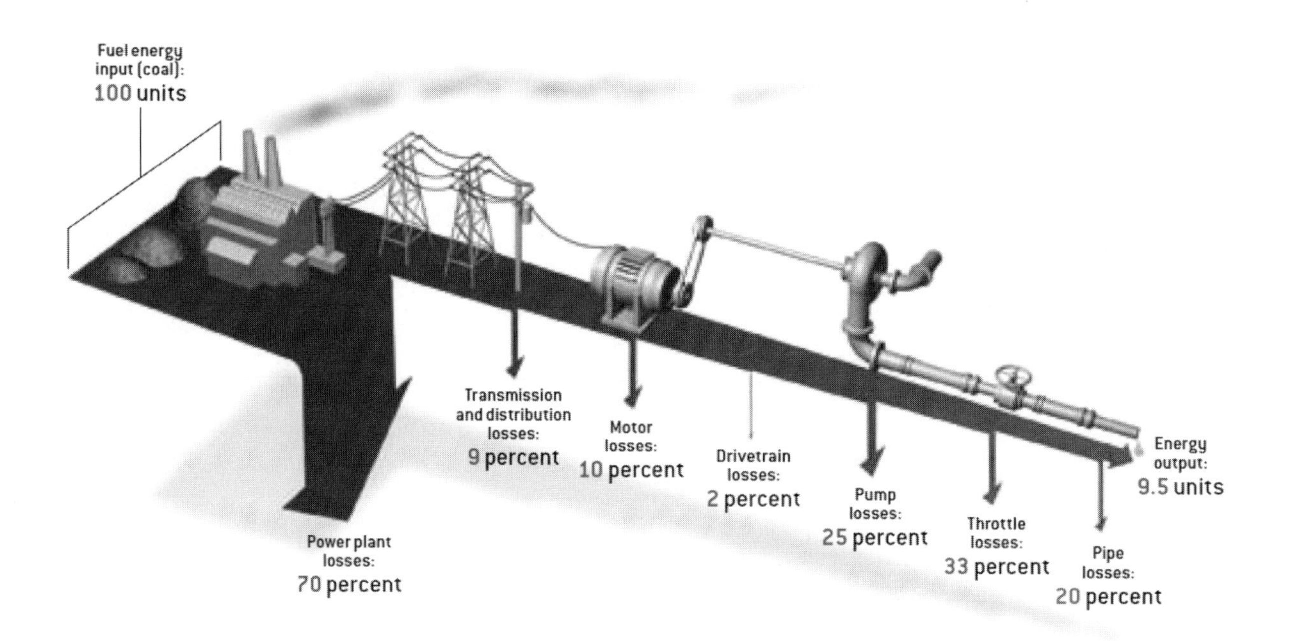

Source: Lovins (2005)[3]

Figure 5.1 *The energy transmission and losses from raw material to the service of a pumped fluid in a typical industrial pumping system*

functional subsystem, contribute very little mass to the PV system. However, the cells' relatively large surface area means the solar module's glass sheets need to be large. Current solar modules also require additional electrical and electronic subsystems such as diodes and inverters. Consequently, most of a PV system's mass is attributed to the supporting subsystems.

In a PV system, solar cells are the subsystem furthest *downstream* in a series of unidirectional subsystem synergies (with respect to both material and energy resources). Figure 5.2 shows the synergies in a PV system. The arrows indicate the dependence between subsystems.

Sliver® cells developed by the Australian National University and Origin Energy are a good example of applying Element 6's features to PV systems. Sliver cells are small, flexible, bifacial solar cells 50–100mm long, 1–2mm wide and 40–70 microns thick. They require about 10 per cent of the silicon[4] (the most costly input in solar-cell production)[5] of conventional cells at a given power rating.

Although Sliver cells have a greater conversion efficiency (over 19 per cent)[6] than conventional solar cells (about 15 per cent), the large majority of the silicon reduction is as a result of the manufacturing technology. About 1000 Sliver cells can be micro-machined from a 150mm, circular, silicon wafer of thickness 1.5mm.[7] The Sliver cells from just one wafer have a total surface area of about 1500cm^2, whereas conventional cells from one wafer have a total surface area of just 177cm^2.[8] The higher cost of complex processing for Sliver cells is offset by the 20 times fewer wafer starts per kW.[9] Sliver cells also have the advantage that they can be made from low-quality or radiation-damaged silicon.[10] Using Sliver cells to build a PV system has many positive *impacts through synergies* on the materials and energy consumption of upstream subsystems. Firstly, Sliver cells' smaller surface area reduces the area and hence mass of the Sliver module's glass sheets. Secondly, Sliver modules can be connected in series without protective diodes and could allow for conversion from DC to AC without a transformer in the inverter.[11] Not only does eliminating these electrical subsystems reduce the PV system's mass, it also improves its energy efficiency, which means the same electricity output can be generated with an even smaller face area.[12] Finally, the Sliver module's lower mass reduces the mass and required integrity of the support structure.

Sliver cells' smaller mass and size also has positive hidden impacts on the mass and energy flow of *upstream subsystems* in the production system (see Figure 5.3). A lower mass requirement for any material also reduces the material and energy demand on the supplier, transporter, processor and extractor.

The lower material demand and energy requirements make Sliver modules at least cost-competitive with conventional solar modules in the retail market, and the energy investment in creating a Sliver module is less than that of a conventional

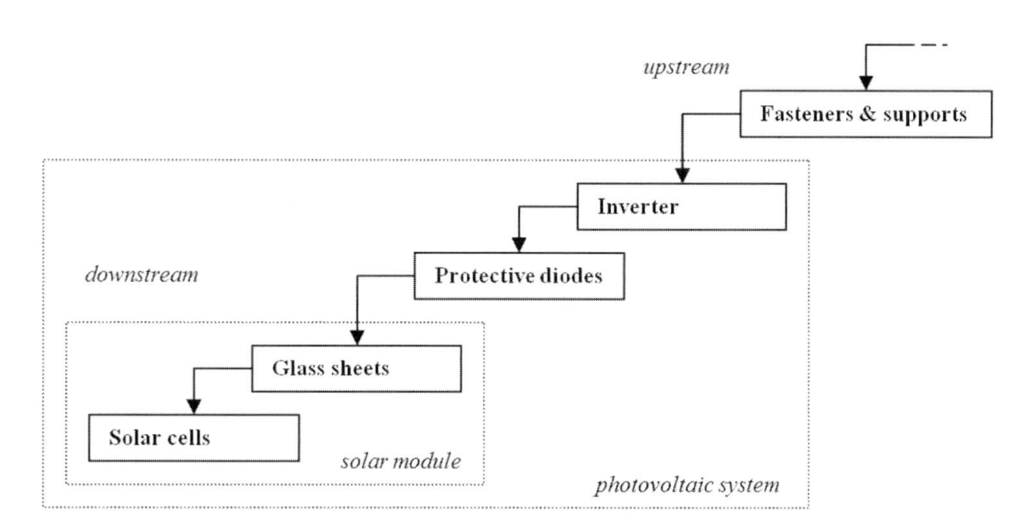

Figure 5.2 *Subsystem synergies in a photovoltaic system with respect to materials and energy resources*

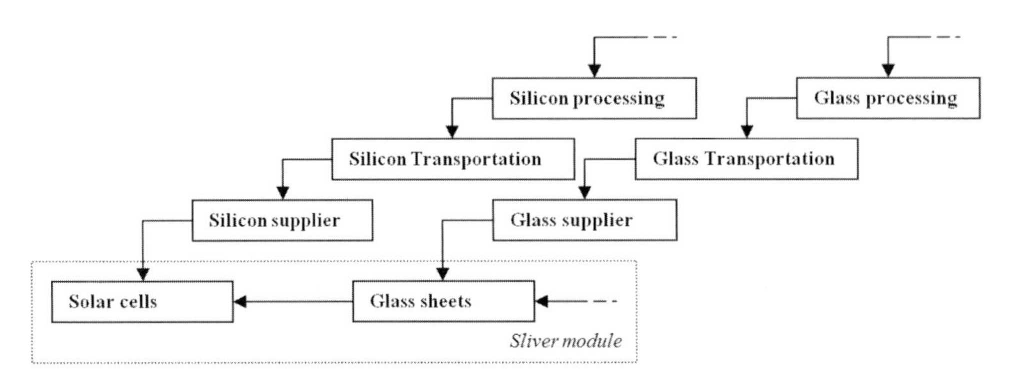

Figure 5.3 *Subsystem synergies in the production system for photovoltaic systems*

module. A Sliver module's energy investment is repaid in 1.5 years of operation when located on a rooftop in a temperate climate.[13] In comparison, the higher silicon content in a conventional module contributes to a much longer, 4.1-year energy payback.[14] Furthermore, the carbon dioxide equivalent coefficient of a Sliver module is about 20 times lower than that of the average electricity generation in Australia (of equivalent generating capacity).[15]

Element 7: Review the system for potential improvements

Chapter 2 highlighted the fact that, in the past, narrow technological solutions often caused more problems than the one they were designed to solve. Examples like leaded petrol and CFCs were given to show just how significant design oversights can be. Hence it is important to *review the whole system to identify potential improvements* to the system's environmental, social and safety performances as well as its cost effectiveness, which is of value primarily during the Detail Design phase. Chapter 2 emphasized that one of the reasons why there have been these major mistakes in the past is a failure to consider impacts of the design on broader systems.

Life-cycle analysis is a tool to help engineers take a precautionary approach with their designs to seek to minimize unforeseen, non-linear, negative system responses, and can help engineers significantly reduce the broader system risks of their design choices.[16] To seek to ensure that such types of design mistakes are not repeated, Systems Engineering (and most engineering disciplines) today recommend using life-cycle analysis databases, chemical-risk databases and occupational health and safety (OH&S) information when assessing design options. The methodology behind Systems Engineering recommends that engineers design with an awareness of the broader systems within which technologies operate – namely the environmental, social and built environment systems. As Blanchard and Fabrycky write:[17]

> Systems engineering involves a life-cycle orientation that addresses all phases to include system design and development, production and/or construction, distribution, operation, maintenance and support, retirement, phase-out, and disposal. Emphasis in the past has been placed primarily on design and system acquisition activities with little (if any) consideration given to their impact on production, operations, maintenance, support and disposal. If one is to adequately identify risks associated with the up-front decision-making process, then such decisions must be based on life-cycle considerations.

Undertaking effective life-cycle analysis can also help to identify new potential sources of energy, water and materials efficiencies. A thorough review of the system to identify potential improvements is at the heart of a WSD approach. As Amory Lovins, Hunter Lovins and Paul Hawken wrote in *Natural Capitalism*:[18]

> At the heart of this chapter, and, for that matter, the entire book, is the thesis that 90-95 per cent reductions in material and energy are possible in developed nations without diminishing the quantity or quality of the services that people want. Sometimes such a large saving can come from a single conceptual or technological leap, like

> Schilham's pumps at Interface in Shanghai ... or a state-of-the-art building. More often, however, it comes from systematically combining a series of successive savings. Often the savings come in different parts of the value chain that stretches from the extraction of a raw resource, through every intermediate step of processing and transportation, to the final delivery of the service (and even beyond to the ultimate recovery of leftover energy and materials). The secret to achieving large savings in such a chain of successive steps is to multiply the savings together, capturing the magic of compounding arithmetic. For example, if a process has ten steps, and you can save 20 per cent in each step without interfering with the others, then you will be left using only 11 per cent of what you started with – an 89 per cent saving overall.

Numerous government energy-efficiency programmes around the world have found there are between 20 and 50 per cent potential energy-efficiency savings across all industry, commercial building and residential sectors. This has been the finding of the Australian Department of Industry, Tourism and Resources (DITR) Energy Efficiency Best Practice programme,[19] which covered a wide range of industry sectors from 1998 to 2003. Through this programme the DITR found that best practice was 80 per cent more energy-efficient than worst practice among dairy processes in Australia, and that even where a company already had an energy-efficiency programme, there were still significant energy-efficiency opportunities to be found.

These results are partly due to the fact that at most sites (from homes to large industrial plants), there is very limited *measurement* and *monitoring* of energy use at the process level. Further, rarely are there properly specified benchmarks against which performance can be evaluated. So rarely do the plant operators know what is possible. Numerous experiences demonstrate that designers and engineers generally assume equipment is working properly when often this is not the case. Lack of measurement, monitoring and benchmarking means that problems can remain undiagnosed for long periods, while wasting energy and money. This can contribute towards risk of failure and increased maintenance costs.

Energy-consuming systems are not often simple. Ideally, they should be *modelled under a range of realistic operating conditions*, so that appropriate priorities for savings measures can be set and reasonable estimates of energy savings from each measure can be made.

Case study: Compressed air system design[20]

Columbia Lighting is a manufacturer of commercial and industrial fluorescent lighting products. One of its plants operates around the clock and has over 300 motors, including a three-motor, 450hp compressed-air system. Fresh out of an electric motor management seminar in 2003, Dennis Short and Scott Patterson of Columbia Lighting were creating a plant-wide inventory of all motors when one of the three motors of the compressed-air system, a 100hp motor, failed. The typical solution was to replace the failed motor with a more efficient model. However, replacing the motor would be twice as expensive as what Columbia Lighting deemed cost-effective.

Instead, Columbia Lighting pursued an alternative two-step solution, which resulted in substantial operating energy and cost reductions. Firstly, they *monitored* for possible air leaks in the compressed-air system using ultrasound techniques on the whole plant. Repairing the leaks reduced energy losses from both pressure drops and heat dissipation, and hence reduced the system's overall power requirement by 47 per cent. Next, they *monitored* the system with the aim of improving the controls. The results were a further 26 per cent reduction in the system's overall power requirement. The overall 73 per cent power savings eliminated the need for replacing the failed motor, since just one of the three original motors, a fixed speed 150hp motor, could now handle the load of the whole compressed-air system.

The improved system has reduced the electrical demand from 152.5kW/h to 37.5kW/h, which translates to an operating cost reduction from US$48,247 per year to US$12,737 per year and a comparable reduction in greenhouse gas emissions – all for the modest cost of assessing and repairing leaks. Further analysis has identified potential energy savings that could make a 100hp variable speed controlled motor a viable option, reducing the overall power requirements for the system by a factor of 4.5 from the original three-motor configuration.

Case study: Addressing fixed energy overheads

It is also important not to overlook the obvious. In most systems – from household appliances to office buildings to industrial sites – the nature of energy use can be characterized as shown in Figure 5.4. In an ideal process, no energy is used when the system is not doing anything useful, but, as Figure 5.4 shows, when this particular industrial plant was not running it was still consuming significant amounts of energy. The gradient of the graph should reflect the ideal amount of energy used to run the process. In practice, most plant and equipment has surprisingly high fixed energy overheads (which could be described as standby energy use). Also the gradient of the typical process is steeper than the ideal graph, reflecting the inefficiencies within the process.

Experience with systems ranging from large industrial plants to retail stores to homes shows similar characteristics. An effective strategy looks at both the fixed energy overheads and the system's marginal efficiency. Often only one or the other is addressed.

Element 8: Model the system

Modelling – including computational/mathematical modelling and computer-aided modelling – is of value during the Need Definition, Conceptual Design, Preliminary Design and Detail Design phases. Field experience, lab tests and computer modelling should be used together where possible to *ensure the system optimum is being approached*, and such techniques are valuable in addressing more complex engineering problems. For instance, the Melbourne University team responsible for successfully redesigning industrial pressurized filtration systems used computer modelling to ensure their redesign was in fact the optimum. Through modelling, the team has been able to improve the efficiency of industrial filtration in existing plants by as much as 40 per cent.

Identifying opportunities for optimization

Modelling by leading Whole System Designers, like RMIT Adjunct Professor Alan Pears, is showing that many everyday products are sub-optimized. Pears's modelling has shown that even the standard dishwasher could be redesigned to no longer need 1.2kWh/wash but instead only 0.56kWh/wash on a normal program.[22] Re-optimizing the system allowed the standard dishwasher to use the least amount of water,

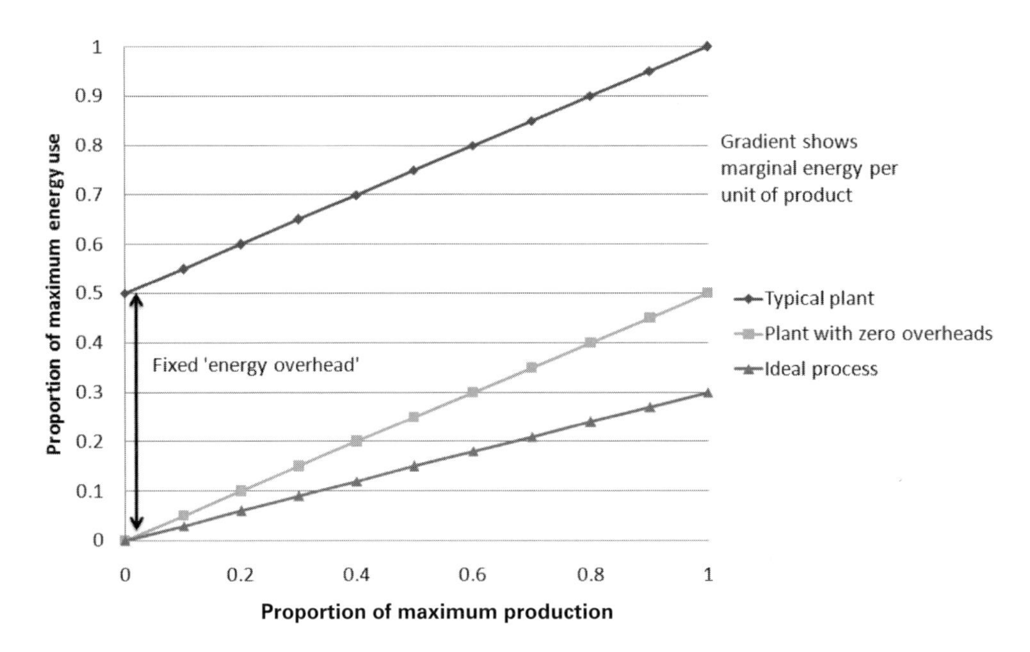

Source: Pears (2004)[21]

Figure 5.4 *Energy use of a typical production system compared with one with zero energy overheads and the ideal process*

operate at the lowest temperatures, and minimize standby electricity consumption and the heat capacity of components heated, while optimizing pump and motor efficiency. Figure 5.5 shows the potential savings the computer modelling identified.

In office buildings in Australia, energy consumption often far exceeds the levels expected on the basis of computer simulation. Thorough inspection and benchmarking usually lead to identification of the reasons for this, and the problems can be rectified. Often the problems are related to relatively minor issues such as inappropriately operated controls, excessive reheat, excessive air leakage into the building and so on.

Genesis Auto, an energy-efficiency consultancy firm led by Geoff Andrews, were contracted to find out why a Commonwealth Government leased building, which had been designed to be 5-star, was only operating at 2–3 star efficiency. To work out why the building was not performing to standard, the building's engineers asked Genesis Auto to meter and monitor the main areas of energy usage – lighting, plug loads (PCs, printers, photocopiers) and the server room. They developed targets for each of these areas of energy usage to ensure that the building achieved a 5-star rating. Metering and monitoring of these three areas concluded that the building was using more energy simply because the plug loads were far higher than anticipated. PCs, printers and photocopiers were being left on all night instead of being turned off. Once this was addressed the building performed at a 5-star level.

Understanding complex systems

Modelling is usually the only cost-effective option for understanding and optimizing complex engineered systems in industry, such as mixing machinery that relies on generating turbulent fluid flows.[24] Conventional 'beat-and-stir' industrial mixing machinery has several practical limitations. Static mixers (which are used in

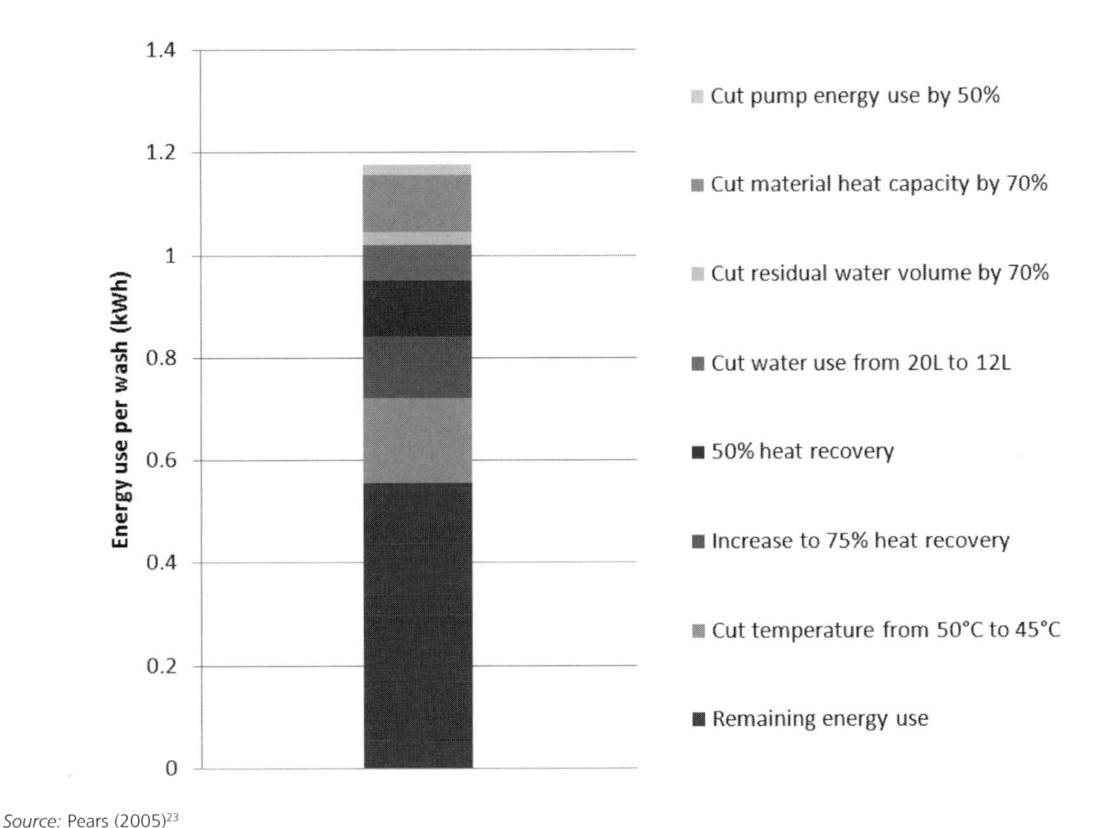

Figure 5.5 *Opportunities to reduce energy consumption in a dishwasher*

cosmetics manufacture) incorporate baffles, plates and constrictions where solids can easily accumulate and impede mixing, thus resulting in a poor product and production downtime. Stirred-tank mixers (which are used in the dairy industry) suffer similar issues. Stirred-tank mixers are also relatively large energy consumers, and they often develop regions of stagnant fluids and regions of high shear, which can result in poor mixing and damage to sensitive and biological materials, and disrupt the formation and growth of particles or aggregates in a crystallizer.

Researchers from CSIRO Energy and Thermofluids Engineering used modelling to develop the rotated arc mixer (RAM). The RAM can mix fluids without the issues of conventional mixers. The RAM relies on very chaotic mixing of highly viscous fluids, where mixing is forced by an outer cylinder rotating around a fixed inner cylinder. The inner cylinder has flow apertures cut at strategic locations, and this configuration creates both axial and transverse flows. The success of mixing is a function of flow rate, rotation rate and flow aperture location. These parameters are optimized using mathematical *modelling*. When the parameters are optimized, the RAM generates very low shear and no stagnant regions, consequently mixing twice as well while consuming five times less energy than a conventional mixer. Modelling is allowing scientists and engineers to develop new designs that significantly reduce environmental impacts.

Case study: Modelling hot-water systems

Imagine if you were given the design challenge to design the most energy- and water-efficient gas hot-water system. What are the right questions to ask? Some of these could include:

- How much hot water is needed by the average household for showers?
- How can the design be optimized to minimize the amount of energy used to heat the water?

Statistics show that the percentage of single-person households in Australia has now increased to 25 per cent of the total, while the percentage of two people per house is also 25. Hence hot-water systems can be designed to meet the needs of just one to two people per house and therefore meet 50 per cent of the residential market for hot-water systems, rather than the current, larger 200-litre systems that are necessary for only 15 per cent of the market. Also the amount of hot water used per household significantly changes if AAA shower heads are used, highlighting the importance of whole-system synergies.

Oversized water heaters have large standby losses. The response of the industry, to date, has been the invention of the instantaneous hot water heater with electronic ignition, which is an improvement. With the instantaneous gas hot-water heater, the water no longer needs to be maintained at a hot temperature all the time. But for 50 per cent of the market in Australia, these heaters are a long way from being optimized for the whole system. Modelling by Adjunct Professor Alan Pears shows that for a household of one to two people, with people having successive showers and with AAA shower heads, a well insulated 30-litre hot-water heater has a large enough capacity to meet their daily shower needs.[25] Modelling by Pears shows that such a highly efficient unit with a well-insulated 30-litre storage tank, using a moderately large burner and electronic ignition, can achieve significantly higher efficiencies than either the 4-star instantaneous hot-water heater systems or the traditional 135-litre systems. As shown in Figure 5.6, this improved efficiency is considerable, right down to very low usage levels. The standby losses are also greatly reduced compared to traditional systems.

In Australia, when the gas industry looks at the performance of their products – in terms of star ratings – they assume a base level of 200 litres is required for showers by the average Australian household. But Pears's analysis of Australian demographics shows that 200l per day is only now needed by 15 per cent of all households. Thus the industry has optimized gas hot-water heaters for an unusual load profile, not an emerging load profile. This highlights the business opportunity for the first company that designs a truly whole-system-optimized gas hot-water system that meets the needs of 50 per cent of Australian households (1–2 person households). Thus it is possible to get a Factor 4 plus improvement through WSD of domestic hot-water systems, and even greater reductions in environmental impact if these insights were used to redesign solar hot-water systems.

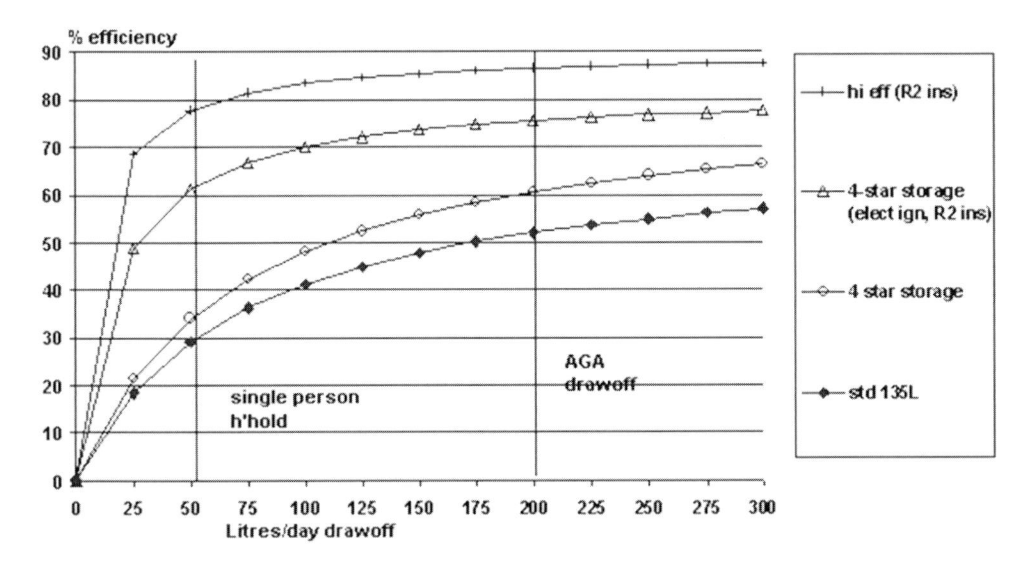

Source: Pears (2003)[26]

Figure 5.6 *Comparison of task efficiencies of standard, 4-star rated and a highly efficient hybrid hot-water system (the significance of managing standby losses is shown by two different options for the 4-star model)*

Element 9: Track technology innovation

One of the main reasons there are still significant resource productivity gains to be made is the fact that *the rate of innovation in basic sciences and technologies has increased dramatically* in the last few decades. Online resources such as Meta-Efficient.com[27] and Engadget.com[28] show that, in many fields, innovations are occurring every six months. *Tracking technology innovation* is of value primarily during the Conceptual Design, Preliminary Design and Detail Design phases. An example of rapid innovation is the refrigerator. The latest innovations in materials science in Europe have created a new insulation material that will allow refrigerators to be 50 per cent more efficient, since most of the energy losses in current systems relate to insulation. This new insulating material achieves R4 levels of insulation while still being very thin. This will enable all heating and cooling devices and appliances, from kettles to microwaves to ovens, to be significantly better insulated without adding significant bulk to the appliance.

Innovations in composite fibres and light metals in materials science now make it possible to design transportation vehicles to be significantly lighter than past car models. Innovations in composite fibres and light metals can now also be used in all forms of transportation, from aircraft to trains to cars, to allow further whole-system improvements.[29] Chapter 7 shows that these materials allow cars to be entirely redesigned.

Innovations in appropriate technology

Innovations in the efficiency of everyday products and renewable energy sources is making the impossible possible. Innovations in ultra-energy-efficient lighting and renewable energy sources now allow many in developing countries to leapfrog the West in terms of energy development. For example, consider that, globally, millions of tons of kerosene,[30] as well as disposable batteries and imported fossil fuels for running small generators, is in widespread use among indigenous populations (who contribute about one third of the world's population). However, these energy sources are relatively costly and the associated technologies are relatively inefficient. The inefficient use of these energy resources creates an opportunity in which the latest energy-efficient and renewable energy technologies can play a significant role in reducing poverty.

This role was recognized in the prestigious journal *Science* in 2005, where Evan Mills, from the US Lawrence Berkeley Labs wrote:

> An emerging opportunity for reducing the global costs and greenhouse gas emissions associated with this highly inefficient form of lighting energy use is to replace fuel-based lamps with white solid-state (LED) lighting, which can be affordably solar-powered. Doing so would allow those without access to electricity in the developing world to affordably leapfrog over the prevailing incandescent and fluorescent lighting technologies in use today throughout the electrified world.[31]

A significant effort is underway to further improve the energy-efficiency of LEDs. LEDs have a market both in developing countries, which can leapfrog current electrical lighting technologies and start using LEDs. In the US, for instance, US$55 billion worth of electricity – some 22 per cent of the nation's total – goes annually to light homes and businesses.

The real advantage here lies in the extremely low maintenance costs due to the low power requirements and long life. Once installed, the powerLED lamps, a type of LED, should last for 20–40 years. The extreme efficiency of LEDs allows them to be powered or to have their batteries recharged through many renewable energy methods – microhydro, wind, solar or biofuels – for low cost. Furthermore, LEDs are available at power ratings as low as 0.5W; the next-lowest-power technology is the compact fluorescent lamp (CFL), the smallest of which is 5W. Even a 5W CLF may be cost-prohibitive in developing countries when powered by a currently costly renewable-energy technology, such as solar photovoltaic panels. The LED's lower power rating allows smaller capacity solar panels to be used and thus helps minimize the total infrastructure cost for modern lighting technologies. Figure 5.7 shows the (a) capital costs and (b) operating costs of various lighting technologies when powered by renewable microhydro technology. Also, due to the kerosene market, there are already distribution networks in place throughout the developing world for LEDs and renewable energy technologies. New organizations, like Lighting Up the World and Barefoot Power, are forming to help both the public and private sectors realize this opportunity. A rapid scaling-up is technically possible and would already be profitable for the private sector. China has the capacity already to produce significant quantities of LEDs very cheaply, and there is enough of a profit margin here for private firms to offer other items of need to villages, such as free malaria nets, as part of the deal to further help address the real causes of poverty, build goodwill and achieve rapid market penetration.

Innovations inspired by nature

Other new areas of innovation come from *biomimicry* – innovation inspired by nature. For the last 300 years, engineers have largely looked to anthropogenic designs and to technical scientific solutions to problems rather than having the humility to learn from nature. CSIRO states that, 'Biomimetic engineering mimics natural systems and processes, using molecular self-assembly as the key link between physics, chemistry and biology, and creating novel advanced structures, materials and devices.'[33] Biomimicry recognizes the fact that the natural world contains highly effective systems and processes which can inform solutions to many of the waste, resource-efficiency and management problems we grapple with today.

Biomimicry has already provided some timely standout innovations in areas such as energy-engineering and waste reuse, where multiple-scale efficiency improvements are greatly needed. Biomimicry's application is predicted across many sectors to help humanity achieve dramatic decoupling of economic growth and negative damage to the environment and communities to create truly restorative systems. Now many scientists and engineers are turning to nature to find new insights into how we can better apply our engineering and design expertise to develop new designs to meet society's needs.

The Natural Edge Project has also developed an introductory training programme on Biomimetic Design for engineers.[34]

Innovations in green chemistry and green engineering

Green chemistry and *green engineering* are remarkable new fields to help engineers achieve Whole System Redesign, right down to the nano level of chemical processes, led by pioneers such as Dr Paul Anastas, Director of the Green Chemistry Institute and former Assistant Director for the Environment in the White

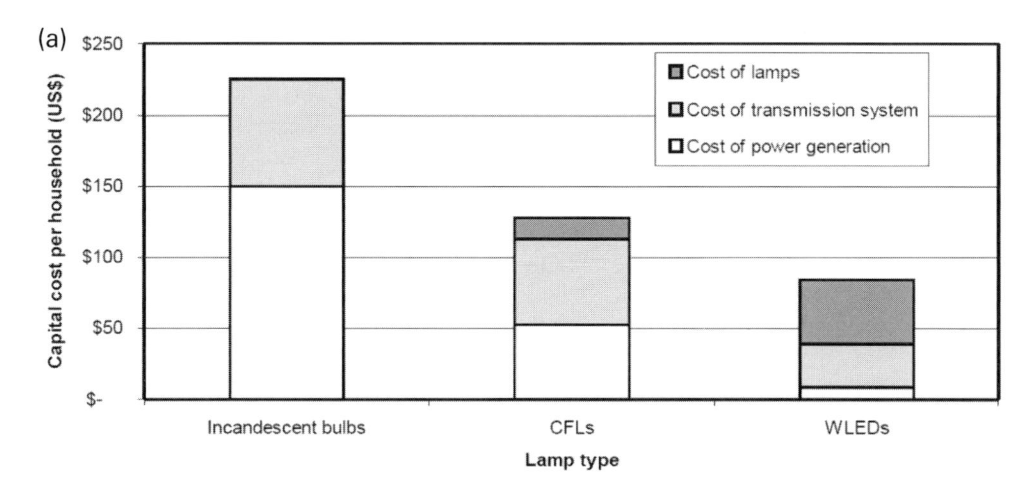

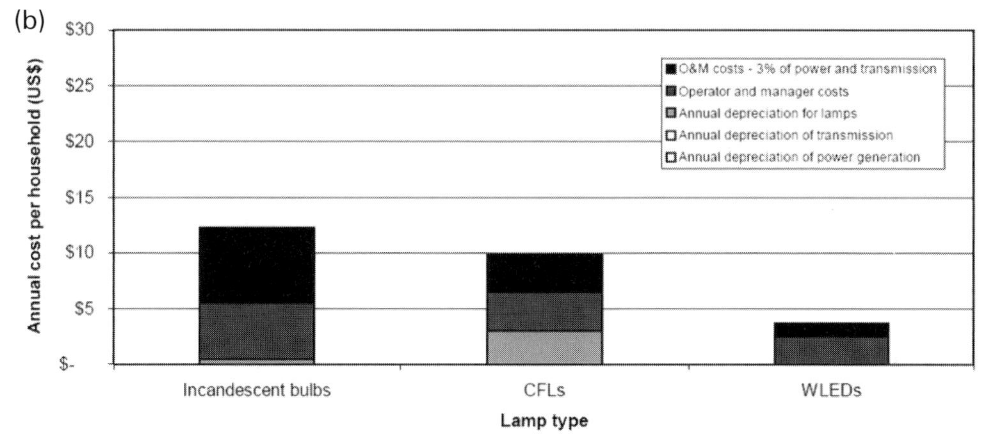

Source: Craine, Lawrance and Irvine-Halliday (2001)[32]

Note: Note that Figures 5.7a and 5.7b were developed in 2001. Since then 1) the cost of CFLs has decreased by 20–50 per cent, 2) the cost of white LEDs (WLEDs) has decreased by about 70 per cent, and 3) the lumens/watt of WLEDs has increased at least two times. In addition, 4) the total lumen output of the WLEDs in Figure 5.7a is about 20 per cent that of the CLFs (although it can be about 50 per cent that of the CLFs in the working area since LED light is focused largely in a single direction). Accounting for these four factors, the capital cost of WLEDs increases to about the same or slightly more than that of CFLs.

Figure 5.7 *Micro hydro village lighting system: Comparison of (a) capital costs and (b) 10-year annual costs per household of various lighting technologies when powered by renewable microhydro technology*

House Office of Science and Technology Policy. Dr Anastas created the 'Green Chemistry Principles', and the new field of knowledge based upon them is helping to guide efforts in the following areas:

- Green chemistry seeks to achieve waste reduction through improved atom economy (that is, reacting as few reagent atoms as possible in order to reduce waste) and reduced use of toxic reagents for the production of environmentally benign products.
- Green chemistry and green chemical engineering seek to utilize catalysts to develop more efficient synthetic routes and reduce waste by avoiding processing steps. Synthetic strategies now employ benign solvent systems (such as ionic water)[35] and supercritical fluids (such as carbon dioxide)[36].

- Biphasic systems and solvent-free methods for many reactions are also being tested to integrate preparation and product recovery. For example, phases of liquids that separate are going to be much easier to recover without needing an additional extractive processing step.
- There has also been significant research into utilizing high-temperature water and microwave heating, sono-chemistry (chemical reactions activated by sonic waves) and combinations of these and other enabling technologies.

Much work is also being done to harness chemicals for common reactions from renewable biomass feedstocks. For instance, in 1989 Szmant estimated that 98 per cent of organic chemicals used in the lab and by industry are derived from petroleum.[37] The Netherlands Sustainable Technology Development[38] project has found that, in principle, there is sufficient biomass production potential to meet the demands for raw organic chemicals from these renewable chemical feedstocks.[39]

An excellent example of green chemistry is the technology developed by Argonne National Lab, a winner of the 1999 US President's Awards for Green Chemistry.[40] Every year in the US alone, an estimated 3.5 million tons of highly toxic, petroleum-based solvents are used as cleaners, degreasers, and ingredients in adhesives, paints, inks and many other applications. More environmentally friendly solvents have existed for years, but their higher costs have kept them from wide use. A technology developed by Argonne National Labs produces non-toxic, environmentally friendly 'green solvents' from renewable carbohydrate feedstocks, such as corn starch. This discovery has the potential to replace around 80 per cent of petroleum-derived cleaners, degreasers, and other toxic and hazardous solvents. The process makes low-cost, high-purity ester-based solvents, such as ethyl lactate, using advanced fermentation, membrane separation and chemical conversion technologies. These processes require very little energy and eliminate the large volumes of waste salts produced by conventional methods. Overall, the process uses as much as 90 per cent less energy and produces ester lactates at about 50 per cent of the cost of conventional methods.

There are currently over 25 research institutions around the globe focused on the development of sustainable chemistry – across Europe, the UK, North America, South America, West Africa and India. The Centre for Green Chemistry[41] in the School of Chemistry at Monash University in Australia is at the forefront of innovation in green chemistry. Established in January 2000, with the goal of providing a fundamental scientific base for future green chemical technology, the centre has a primary focus on Australian industry and Australian environmental problems. Among emerging green chemistry centres worldwide, it is noteworthy for its broad spectrum of research interests, including benign technologies for corrosion inhibitors, gold processing and greener reaction media for chemical synthesis, to name but a few. Green Chemistry Principles, as pioneered by Dr Anastas, and the field of knowledge that is growing based upon them are helping to guide chemists and chemical engineers in their efforts to assist industry in its drive towards sustainability.

The Natural Edge Project has also developed an introductory training programme on Green Chemistry and Green Engineering.[42]

Element 10: Design to create future options

A basic tenet of sustainability is that future generations should have the same level of life quality, environmental amenities and range of options as 'developed' societies enjoy today. Chapters 6–10 will provide examples of designing systems that can aid society in its transition to sustainability. However, it is also important to consider going beyond best practice by helping to create more options for future generations, as shown in Figure 5.8.

Designing to create options is not an abstract idea. It is crucial that today's designers are aware of how their systems affect the options of future generations. For example, as we discussed at the end of Chapter 3, China is currently developing new coal-fired power stations at a rate of one per week. However, it is vital that new coal-fired power stations can be used for geo-sequestration when the technology becomes commercially available. There are significant concerns that many coal-fired power stations in development are not correctly sited nor designed to make geo-sequestration of CO_2 emissions possible in the future. To further demonstrate this element, consider the following examples:

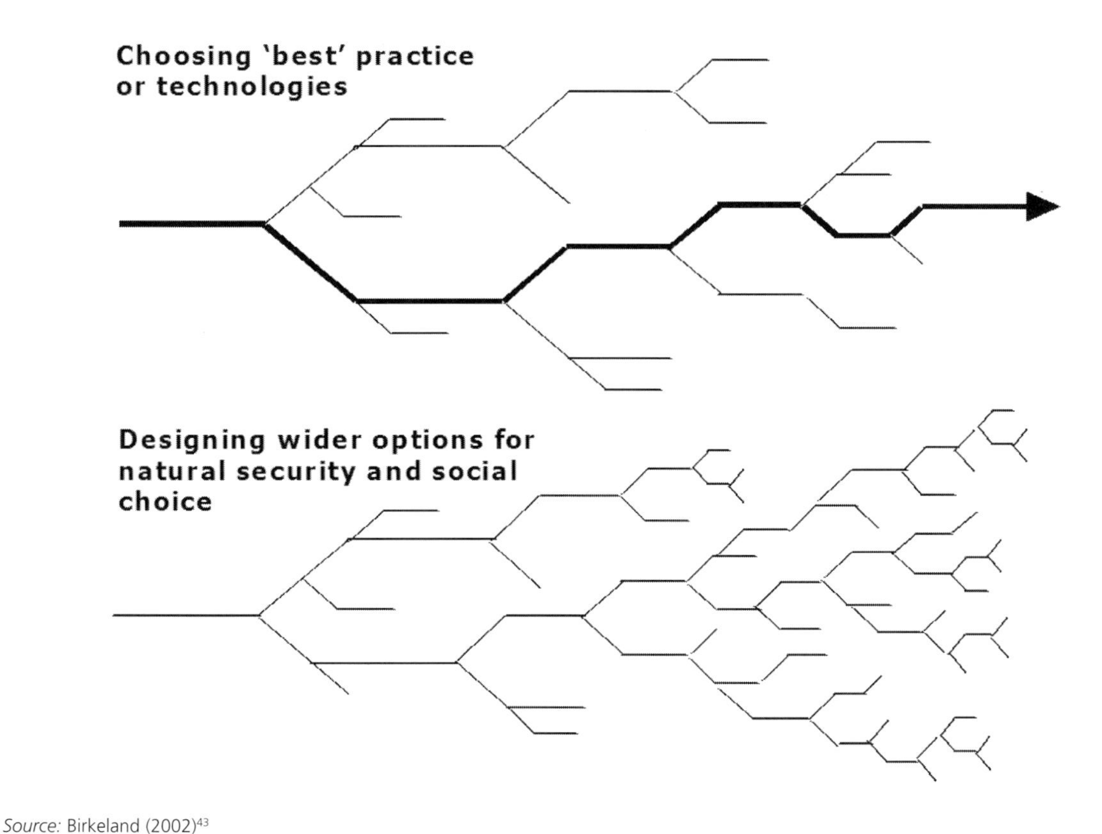

Source: Birkeland (2002)[43]

Figure 5.8 *The standard decision tree compared to a sustainability design tree*

- *Pipes and pumping systems* (covered in Chapter 6): This worked example shows that it is possible to reduce the negative impact on the environment by up to 90 per cent. In some cases, there is additional opportunity to design pipes that give future generations more options. For example, in China new gas pipelines are designed to also accommodate hydrogen in the future.
- *Hybrid cars* (covered in Chapter 7): This worked example shows that it is possible to significantly improve the fuel-efficiency of cars, which then opens up new fuel options. Improved fuel-efficiency makes biofuels and hydrogen fuel sources cost-effective. General Motors' new plug in hybrid car concept car eFlex is designed to run on petrol, biofuels or hydrogen, ensuring that the car design can take advantage of whichever fuel mixes dominate the market in the future. General Motors' Head of Development, Jon Lauckner, has committed to producing the world's first commercial plug-in hybrid. Car designers are also trying to improve the array of options for future generations by designing cars and their electrical components to be over 90 per cent re-manufacturable. Remanufacturability is now a requirement in many countries in Europe and Asia, where the manufacturer's responsibility for its products is being extended to the entire life-cycle (see the featured Hypercar Revolution case study below.)
- *The IT and electronics industry* (covered in Chapter 8): This worked example shows that a WSD approach to server design can greatly reduce energy consumption. IT must also be designed for remanufacture and recycling, which can reduce e-waste and ensure that precious metals and resources can be reused. The Natural Edge Project has also developed an introductory training programme on e-waste.[44]

- *The building industry* (covered in Chapter 9): This worked example shows that a WSD approach to building design can reduce energy consumption. Many designers are also developing buildings where the materials can be dismantled and reused, such as the award-winning Newcastle University green buildings.
- *Domestic water systems* (covered in Chapter 10): This worked example shows that a WSD approach to water-consuming systems in the home can greatly reduce water consumption. Beyond the scope of this worked example, dual pipes are a requirement for new building developments in many countries, so that future occupants can choose to reuse their grey water.

There are a number of tools to assist designers to design for increased choice for future generations, such as backcasting and design or end-of-life processing.

Backcasting from the future system

Backcasting involves designing a 'future system', a system for an envisioned future, by considering desired technological and political states, and then working backwards to develop a system that most closely matches that future system with technologies and policies that are available now.[45] The envisioned future should represent the *desired outcome* rather than the transition,[46] and should be general and *non-prescriptive* so as to be applicable at many levels, to many fields and to many industries. A general vision will encourage a flexible system that can adapt to unforeseen *technological* and *political* disturbances. Backcasting is of value primarily during the Conceptual Design and Preliminary Design phases.

Table 5.1 contrasts forecasting and backcasting. In many ways, forecast systems emulate systems that are backcast, but from only a short time into the future.

In practice, the technical inadequacies of forecasting are usually exacerbated by favouring the short-term 'safe bet' option. Indeed, many modern 'innovative' systems based on forecasting still reflect a compromise between the best (long-term) option and the risk of a costly failed venture. Figure 5.9 compares forecasting and backcasting using an elastic band analogy (wherein the original system is akin to a slightly taut elastic band around two pegs).

Case study: Passenger vehicle design

The basic structure of the car has changed very little in the past 50 years and is based on a design platform that first appeared about 100 years ago, at which time that platform was probably *optimal*. The *state of technology* at that time suggested that medium and long-distance mobility could be most efficiently and cheaply achieved by a system such as Henry Ford's Model T – the first mass-produced, petrol-driven vehicle with a transmission mechanism. Since then, and particularly in the last 50 years, cars have evolved *incrementally* and *cautiously*, while technology has progressed rapidly. Consequently, modern passenger vehicles could be substantially different from existing cars and could potentially be more *optimal*. Hypercar used backcasting to develop a new passenger vehicle platform.[48] The new platform is optimal in an *envisioned future of technological and political sustainability* and has the characteristics listed in Figure 5.10. The modern-day passenger vehicle of Hypercar's platform is the Hypercar Revolution. Several subsystems of the Revolution, including the advanced composite structure, hydrogen fuel-cells, the Fiberforge™ manufacturing process, and extensive electronic and software control are a result of backcasting from a future system (see Figure 5.10). These subsystems are currently viable steps on the path to a sustainable vehicle.

The Revolution outperforms benchmark vehicles in every category measured. However, it is a concept

Table 5.1 *Contrasting conventional forecasting and backcasting*

Forecasting	Backcasting
Is influenced by the *current* technological and political states, and...	Focuses on an envisioned future with *desired* technological and political states, which...
often only looks as far as the *next* improvement; thus...	is *independent* of both time and current technological and political states; thus...
facilitates the *propagation* of otherwise out-dated trends.	facilitates the *termination* of out-dated trends.
Is more of a market *push* based on *convenience* and *security*.	Is more of a market *pull* based on *need* and *incentive*.

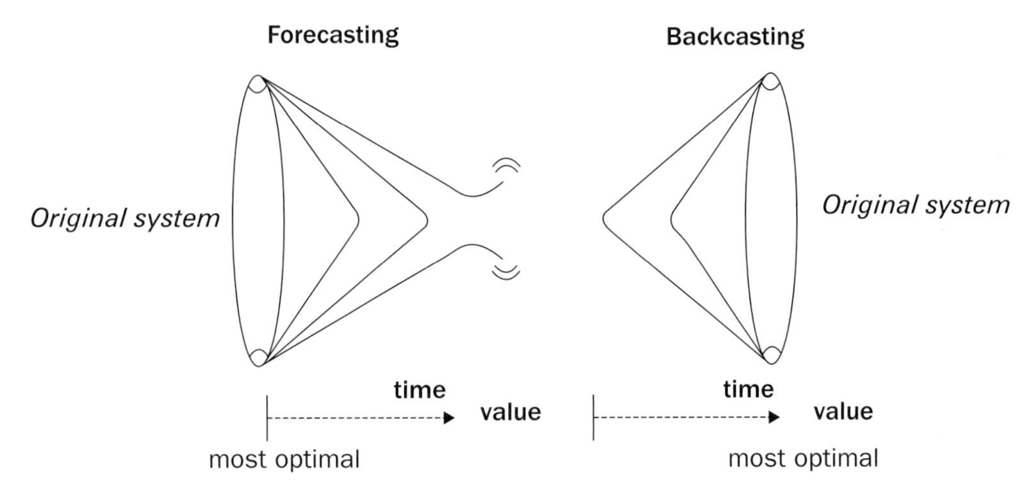

Source: Adapted from Lovins (2002)[47]

Forecasting: The original system is the first and worst of the series but is also the most optimal for current technological and political states. Forecasting introduces a fundamental design compromise. Every forecast upgrade pulls the system away from its original design (original elastic position), but since the platform does not change with technological and political progress (the pegs are fixed), the ease of adding value to a system (by pulling away further) reduces with time. This effect is known as diminishing returns.
Backcasting: The original system is the last and best model of the series, and is also the most optimal for future technological and political states. Every backcast upgrade pulls the system towards its original design, and since this is in the direction of technological and economic progress, the ease of adding value to a system increases with time. This effect is known as expanding returns.

Figure 5.9 *Using the elastic band analogy to compare forecasting with backcasting*

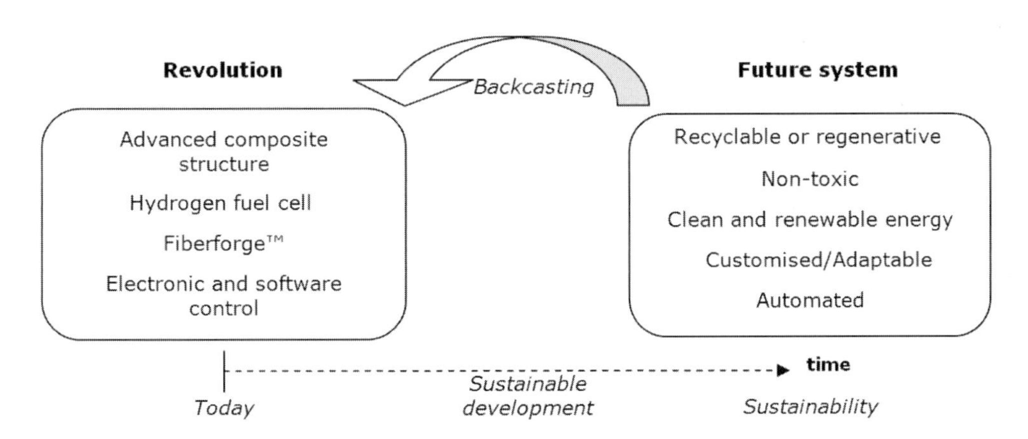

Figure 5.10 *Backcasting a sustainable passenger vehicle platform*

vehicle and, although it can be competitively mass-produced, support infrastructure and resources, such as refuelling stations for hydrogen fuel-cell technology, are not yet sufficient. In this situation it is worthwhile backcasting further to develop a vehicle wherein the fuel-cell-electric power plant is preceded by a petrol-electric power plant. Since the rest of the vehicle is compatible with a fuel-cell-electric power plant, a future upgrade would be relatively cheap and easy.

Design for end-of-life processing

Design for end-of-life has a large influence on the system's legacy and is of value primarily during the Detail Design phase. Good design seeks to ensure that non-biological resources can be reclaimed easily at end-of-life, or returned to a 'technological metabolism', as Bill McDonough famously put it. This also helps make companies and customers money by optimizing the system's salvage value. Designing for end-of-life processing involves the following:[49]

- *Ease of disassembly*: Where disassembly cannot be avoided, making it easier can reduce the time required for this non-value-adding activity. Permanent fastening such as welding or crimping should not be used if the product is intended for remanufacture. Components should not be damaged during disassembly.
- *Ease of cleaning*: Used components will likely require cleaning. Design components for easy cleaning by understanding the cleaning methods, making the surfaces to be cleaned accessible and ensuring that cleaning residues cannot accumulate on the component.
- *Ease of inspection*: Minimize the time required for this non-value-adding activity.
- *Ease of part replacement*: Components that wear should be easily accessible so as to minimize the time required for reassembly and prevent damage during component insertion.
- *Ease of reassembly*: Minimize the time required for reassembly – if the system is being processed at end-of-life, then it will be assembled multiple times throughout its life. Be aware of tolerances between components.
- *Reusable components*: Increasing the number or reusable components increases the cost effectiveness of end-of-life processing.
- *Modular components*: Modular systems require less time for assembly and disassembly.
- *Fasteners*: Using fewer different fasteners reduces the complexity of assembly, disassembly and the materials handling.
- *Interfaces*: Using fewer different component interfaces reduces the number of different components required to produce a family of systems, which helps build economies of scale and improve re-manufacturability.

In the 1970s and 1980s, Walter Stahel and colleagues proposed that redesigning products to minimize waste, resources and energy was a good place to start to achieve a sustainable society. In 1982, they formed the Product Life Institute in Geneva to further these studies.[50] They developed for the first time the methodologies for many of the strategies now accepted today, such as *extended product responsibility*. They developed the ideas of how society needs to shift from a linear 'cradle to grave' approach to a cyclical 'cradle to cradle' approach for product design and use to minimize waste. Stahel[51] and colleagues pioneered the concepts behind 'cradle to cradle', arguing that the following would help to achieve it:[52]

- Product design should be optimized for durability, remanufacturing and recycling;
- Remanufacturing – preserving the stable frame of a product after use and replacing only worn out parts;
- Leasing instead of selling[53] – wherein the manufacturer's interest lies in durability; and
- Extended Product Liability/Stewardship/Responsibility – which could induce manufacturers to guarantee low pollution use and easy reuse.

Walter Stahel proposed three basic approaches to encouraging the reduction and minimization of waste. These are outlined in detail in his Mitchell Prize Award-winning essay, '*The product-life factor*',[54] in which Stahel proposed a complex product life extension system (see Figure 5.11):

> A Self-Replenishing System would create an economy based on a spiral loop system that minimizes matter, energy-flow and environmental deterioration without restricting economic growth or social and technical progress.

Stahel's 1982 diagram (Figure 5.11) described how, through reuse (loop 1), repair (loop 2) and reconditioning (loop 3), it is possible to utilize used products or components as a source for new ones, as well as recycling (loop 4), using scrap as locally available raw material.

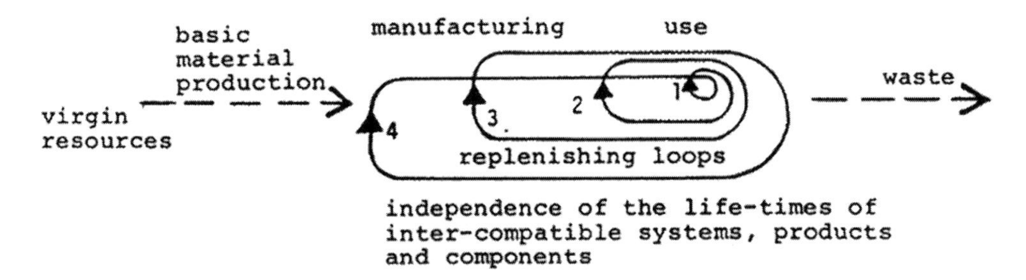

Source: Stahel, W.R. (1982)[55]

Figure 5.11 *The Self-Replenishing System (product life extension)*

Optional reading

Amezquita, T., Hammond, R., Salazar, M. and Bras, B. (1995) 'Characterising the remanufacturability of engineering systems', proceedings of ASME Advances in Design Automation Conference, Boston, MA, vol 82, pp271-278, www.srl.gatech.edu/education/ME4171/DETC95_Amezquita.pdf, accessed 29 July 2007

Birkeland, J. (2002) *Design for Sustainability: A Sourcebook of Ecological Design Solutions*, Earthscan, London

Çengel, Y. and Boles, M. (2008) *Thermodynamics – An Engineering Approach* (6th edition), McGraw-Hill, pp78-96, http://highered.mcgraw-hill.com/classware/infoCenter.do?isbn=0073529214&navclick=true, accessed 28 August 2008

Hawken, P., Lovins, A. B. and Lovins, L. H. (1999) *Natural Capitalism: Creating the Next Industrial Revolution*, Earthscan, London, Chapter 6: 'Tunnelling through the cost barrier', www.natcap.org/images/other/ NCchapter6.pdf, accessed 26 July 2007

Lovins, A. B., Datta, E. K., Bustnes, O. E., Koomey, J. G. and Glasgow, N. J. (2004) *Winning the Oil Endgame: Innovation for Profits, Jobs and Security, Technical Annex*, Rocky Mountain Institute, Snowmass, CO, www.oilendgame.com/TechAnnex.html, accessed 29 July 2007

McDonough, W. and Braungart, M. (2002) *Cradle to Cradle: Remaking the Way We Make Things*, North Point Press, New York

Pears, A. (2004) 'Energy efficiency – Its potential: Some perspectives and experiences', background paper for International Energy Agency Energy Efficiency Workshop, Paris, April, www.naturaledgeproject.net/Documents/IEAENEFFICbackgroundpaperPears Final.pdf, accessed 29 July 2007

Romm, J. J. and Browning, W. D. (1998) *Greening the Building and the Bottom Line*, Rocky Mountain Institute, CO

Stahel, W. R. (1982) *The Product Life Factor*, The Product-Life Institute, Geneva, www.product-life.org/milestone2.htm, accessed 29 July 2007 (this was a Mitchell Prize-Winning paper)

Van der Ryn, S. and Cowan, S. (1995) *Ecological Design*, Island Press

Van der Ryn, S. (2005) *Design for Life: The Architecture of Sim Van der Ryn*, Gibbs-Smith Publishers, http://64.143.175.55/va/index-methods.html, accessed 29 July 2007

Von Weizsäcker, E., Lovins, A. B. and Lovins, L. H. (1997) *Factor Four: Doubling Wealth, Halving Resource Use*, Earthscan, London

William McDonough Architects (1992) *Hanover Principles of Design for Sustainability*, prepared for EXPO 2000, The World's Fair, Hanover, Germany, www.mcdonough.com/principles.pdf, accessed 14 August 2007

Notes

1 Hawken, P., Lovins, A. B. and Lovins, L. H. (1999) *Natural Capitalism: Creating the Next Industrial Revolution*, Earthscan, London, p115.

2 Lamb, G. (2005) 'User's guide to pump selection', *WME Magazine*, July, pp40–41.

3 Lovins, A. B. (2005) 'More profit with less carbon', *Scientific American*, September, p76, www.sciam.com/media/pdf/Lovinsforweb.pdf, accessed 11 April 2008.

4 Blakers, A., Weber, K., Everett, V., Deenapanray, S. and Franklin, E. (2004) *Sliver Solar Cells and Modules*, 42nd Annual Conference of the Australian and New Zealand Solar Energy Society, http://energy.murdoch.edu.au/Solar2004/Proceedings/Photovoltaics/Blakers_Paper_Silver.pdf, accessed 26 March 2005.

5 Blakers, A. and Stock, A. (2002) *New Sliver Cell Offers Revolution in Solar Power*, Origin Energy,

www.originenergy.com.au/news/news_detail.php?newsid=233&pageid=82, accessed 10 April 2005.

6 Blakers, A., Weber, K., Everett, V., Deenapanray, S. and Franklin, E. (2004) *Sliver Solar Cells and Modules*, 42nd Annual Conference of the Australian and New Zealand Solar Energy Society, http://energy.murdoch.edu.au/Solar2004/Proceedings/Photovoltaics/Blakers_Paper_Silver.pdf, accessed 7 June 2007.

7 Blakers, A., Weber, K., Everett, V., Deenapanray, S. and Franklin, E. (2004) *Sliver Solar Cells and Modules*, 42nd Annual Conference of the Australian and New Zealand Solar Energy Society, http://energy.murdoch.edu.au/Solar2004/Proceedings/Photovoltaics/Blakers_Paper_Silver.pdf, accessed 7 June 2007.

8 Blakers, A., Weber, K., Everett, V., Deenapanray, S. and Franklin, E. (2004) *Sliver Solar Cells and Modules*, 42nd Annual Conference of the Australian and New Zealand Solar Energy Society, http://energy.murdoch.edu.au/Solar2004/Proceedings/Photovoltaics/Blakers_Paper_Silver.pdf, accessed 7 June 2007.

9 Blakers, A., Weber, K., Everett, V., Deenapanray, S. and Franklin, E. (2004) *Sliver Solar Cells and Modules*, 42nd Annual Conference of the Australian and New Zealand Solar Energy Society, http://energy.murdoch.edu.au/Solar2004/Proceedings/Photovoltaics/Blakers_Paper_Silver.pdf, accessed 7 June 2007.

10 Blakers, A., Weber, K., Everett, V., Deenapanray, S. and Franklin, E. (2004) *Sliver Solar Cells and Modules*, 42nd Annual Conference of the Australian and New Zealand Solar Energy Society, http://energy.murdoch.edu.au/Solar2004/Proceedings/Photovoltaics/Blakers_Paper_Silver.pdf, accessed 7 June 2007.

11 Stocks, M. J. et al (2003) '65-micron thin monocrystalline silicon solar cell technology allowing 12-times reduction in silicon usage', paper presented at 3rd World Conference of Photovoltaic Solar Energy Conversion, Osaka, Japan, p3, http://solar.anu.edu.au/docs/65micronthinmonosi.pdf, accessed 7 June 2007.

12 Duffin, M. (2004) 'The energy challenge 2004: Solar', *EnergyPulse*, www.energypulse.net/centers/article/article_display.cfm?a_id=864, accessed 7 June 2007.

13 Deenapanray, P. N. K., Blakers, A. W., Weber, K. J. and Everett, V. (2004) *Embodied Energy of Sliver Modules*, 19th European PV Solar Energy Conference, http://solar.anu.edu.au/level_1/pubs/papers/2CV_3_35.pdf, accessed 7 June 2007.

14 Deenapanray, P. N. K., Blakers, A. W., Weber, K. J. and Everett, V. (2004) *Embodied Energy of Sliver Modules*, 19th European PV Solar Energy Conference, http://solar.anu.edu.au/level_1/pubs/papers/2CV_3_35.pdf, accessed 7 June 2007.

15 Deenapanray, P. N. K., Blakers, A. W., Weber, K. J. and Everett, V. (2004) *Embodied Energy of Sliver Modules*, 19th European PV Solar Energy Conference, http://solar.anu.edu.au/level_1/pubs/papers/2CV_3_35.pdf, accessed 7 June 2007.

16 See further explanation on the role of engineers in designing benign technical solutions in Smith, M., Hargroves, K., Paten, C. and Palousis, N. (2007) *Engineering Sustainable Solutions Program: Critical Literacies Portfolio - Principles and Practices in Sustainable Development for the Engineering and Built Environment Professions*, The Natural Edge Project, Australia, www.naturaledgeproject.net/TNEP_ESSP_CLP_Principles_and_Practices_in_Sustainable_Development_for_the_Engineering_and_Built_Environment_Professions.aspx, accessed 3 July 2007.

17 Blanchard, B. S. and Fabrycky, W. J. (2006) *Systems Engineering and Analysis* (fourth edition), Pearson Prentice Hall, Upper Saddle River, NJ, Chapter 1.

18 Hawken, P., Lovins, A. B. and Lovins, L. H. (1999) *Natural Capitalism: Creating the Next Industrial Revolution*, Earthscan, London.

19 See Department of Industry, Tourism and Resources, 'Energy Efficiency Best Practice programme' at www.ret.gov.au/Programsandservices/EnergyEfficiencyBestPracticeEEBPProgram/Pages/default.aspx, accessed 12 May 2007.

20 Electric Motor Management (2004) 'Motor management success: Information, cooperation and teamwork lead to superior decisions at Columbia Lighting', Electric Motor Management, www.drivesandmotors.com/downloads/Columbia_SS_Final.pdf, accessed 7 March 2006.

21 Pears, A. (2004) 'Energy efficiency - Its potential: Some perspectives and experiences', background paper for International Energy Agency Energy Efficiency Workshop, Paris, p12, www.naturaledgeproject.net/Documents/IEAENEFFICbackgroundpaperPearsFinal.pdf, accessed 30 March 2008.

22 Pears, A. (2005) 'Design for energy efficiency', presentation to Young Engineers Tasmania; private communication.

23 Pears, A. (2005) 'Design for energy efficiency', presentation to Young Engineers Tasmania; private communication.

24 See CSIRO, 'Revolutionary new mixer mixes the unmixable' at www.cmit.csiro.au/brochures/serv/ram/, accessed 7 June 2007.

25 But if it is assumed that the household is not using efficient AAA shower heads and instead are using inefficient shower heads, then 30 litres will not be enough.

26 Pears, A. (2003) *Household Hot Water and Sustainability – What's Wrong with Existing Technologies and How to Fix Them*, RMIT, Australia.

27 See Meta-Efficient website at www.metaefficient.com, accessed 7 June 2007.

28 See Engadget website at www.engadget.com, accessed 7 June 2007.

29 Lovins, A. B., Datta, E. K., Bustnes, O. E., Koomey, J. G. and Glasgow, N. J. (2004) *Winning the Oil Endgame: Innovation for Profits, Jobs and Security, Technical Annex*, Rocky Mountain Institute, Snowmass, CO, www.oilendgame.com/TechAnnex.html, accessed 29 July 2007.

30 Mills, E. (2000) *Global Lighting Energy Use and Greenhouse Gas Emissions*, Lawrence Berkley Laboratories, US.

31 Mills, E. (2005) 'The specter of fuel-based lighting', *Science*, no 308, pp1263-1264, www.sciencemag.org/cgi/content/summary/308/5726/1263, accessed 29 July 2007.

32 Craine, S. and Irvine-Halliday, D. (2001) 'White LEDs for lighting remote communities in developing countries', in (eds) I. T. Ferguson, Y. S. Park, N. Narendran and S. P. DenBaars (eds) *Solid State Lighting and Displays*, Proceedings of SPIE (Society of Photo-Optical Instrumentation Engineers), vol 4445, pp39–48.

33 CSIRO (2007) *New Membrane Materials: Biomimetics*, CSIRO, www.csiro.au/science/ps30k.html, accessed 28 August 2007.

34 Smith, M., Hargroves, K., Desha, C. and Palousis, N. (2007) *Engineering Sustainable Solutions Program: Critical Literacies Portfolio – Principles and Practices in Sustainable Development for the Engineering and Built Environment Professions*, The Natural Edge Project (TNEP), Australia, Unit 3, Lectures 9 and 10, www.naturaledgeproject.net/ESSPCLP-Principles_and_Practices_in_SD-Lecture9.aspx, accessed 29 July 2007.

35 Breslow, R. (1998) 'Water as a solvent for chemical reactions', in P. Anastas and T. Williamson (eds) *Green Chemistry, Frontiers in Design Chemical Synthesis and Processes*, Oxford University Press; Li, C. (2000) 'Water as solvent for organic and material synthesis', in P. Anastas, L. Heine, T. Williamson and L. Bartlett (eds) *Green Engineering*, American Chemical Society.

36 Hancu, D., Powell, C. and Beckma, E. (2000) 'Combined reaction-separation processes in CO_2', in P. Anastas, L. Heine, T. Williamson and L. Bartlett (2000) *Green Engineering*, American Chemical Society.

37 Szmant, H. (1989) *Organic Building Blocks of the Chemical Industry*, Wiley, New York, p4.

38 Weaver, P., Jansen, J., van Grootveld, G., van Spiegel, E. and Vergragt, P. (2000) *Sustainable Technology Development*, Greenleaf Publishers, Sheffield, UK.

39 Okkerse, C. and Van Bekkum, H. (1996) 'Renewable raw materials for the chemicals industry', in *Sustainability and Chemistry*, Sustainable Technology Development, Delft, The Netherlands.

40 Argonne National Lab (1998) 'Green solvent process gets Presidential honor', *Argonne News*, www.anl.gov/Media_Center/Argonne_News/news98/an980629.html, accessed 28 August 2007; Argonne National Lab (1999) 'Green solvent process wins federal award', *Argonne News*, 1 March, www.anl.gov/Media_Center/Argonne_News/news99/an990301.html, accessed 28 August 2007.

41 See 'The Centre for Green Chemistry' on the School of Chemistry at Monash University website at www.chem.monash.edu.au/green-chem/, accessed 29 July 2007.

42 Smith, M., Hargroves, K., Desha, C. and Palousis, N. (2007) *Engineering Sustainable Solutions Program: Critical Literacies Portfolio – Principles and Practices in Sustainable Development for the Engineering and Built Environment Professions*, The Natural Edge Project (TNEP), Australia, Unit 3, Lectures 11 and 12, www.naturaledgeproject.net/ESSPCLP-Principles_and_Practices_in_SD-Lecture11.aspx, accessed 29 July 2007.

43 Birkeland, J. (2002) 'Unit notes', University of Canberra.

44 Hargroves, K., Stasinopoulos, P., Desha, C. and Smith, M. (2007) *E-Waste Education Courses*, The Natural Edge Project, Australia, www.naturaledgeproject.net/EWasteHome.aspx, accessed 7 July 2007.

45 Holmberg, J. and Robert, K. H. (2000) 'Backcasting from non-overlapping sustainability precepts: A framework for strategic planning', *International Journal of Sustainable Development and World Ecology*, no 7, pp291–308.

46 Holmberg, J. and Robert, K. H. (2000) 'Backcasting from non-overlapping sustainability precepts: A framework for strategic planning', *International Journal of Sustainable Development and World Ecology*, no 7, pp291–308.

47 Adapted from Lovins, A. B. (2002) *FreedomCAR, Hypercar and Hydrogen*, Rocky Mountain Institute, CO, www.rmi.org/images/other/Trans/T02-06_FreedomCAR.pdf, accessed 17 January 2007.

48 Adapted from Lovins, A. B. (2002) *FreedomCAR, Hypercar and Hydrogen*, Rocky Mountain Institute, CO, www.rmi.org/images/other/Trans/T02-06_FreedomCAR.pdf, accessed 17 January 2007.

49 Amezquita, T., Hammond, R., Salazar, M. and Bras, B. (1995) *Characterising the Remanufacturability of Engineering Systems*, proceedings 1995 ASME Advances in Design Automation Conference, Boston, MA, vol 82, pp271–278, www.srl.gatech.edu/education/ME4171/DETC95_Amezquita.pdf, accessed 16 May 2007.

50 See Product Life Institute website at www.product-life.org, accessed 26 November 2006.

51 Stahel, W. R. (1982) *The Product-Life Factor*, The Product-Life Institute, Geneva, www.product-life.org/

milestone2.htm, accessed 30 July 2007. This was a Mitchell Prize-winning paper.

52 Stahel, W. and Gomvingen, E. (1993) *Gemeinsam nutzen statt einsela verbrauchen*, International Design Forum/IFG, Giessen, Germany, Anabas Verlag.

53 Beginning in the mid-1980s, Swiss industry analyst Walter Stahel and German chemist Michael Braungart independently proposed a new industrial model that is now gradually taking shape. Rather than an economy in which goods are made and sold, these visionaries imagined a service economy wherein consumers obtain services by leasing or renting goods rather than buying them outright.

54 Stahel, W. R. (1982) *The Product-Life Factor*, The Product-Life Institute, Geneva, www.product-life.org/milestone2.htm, accessed 30 July 2007. This was a Mitchell Prize-winning paper.

55 Stahel, W. R. and Reday, G. (1976) 'The potential for substituting manpower for energy', report to the European Commission, Brussels, published 1982 as *Jobs for Tomorrow – The Potential for Substituting Manpower for Energy*, Vantage Press, New York.

6

Worked Example 1 – Industrial Pumping Systems

Significance of pumping systems and design

Motors use 60 per cent of the world's electricity, and of this percentage, 20 per cent is used for pumping.[1] Such a large portion is no surprise, as most systems are running in continuous operation for 18 hours per day or more.

With such a large amount of energy devoted to moving liquid from one place to another (a lot of which is used to fight pipe friction and in many cases unnecessary changes in height and direction), improving the efficiency of industrial pumping systems can make major strides in the reduction of industrial energy consumption and hence greenhouse emissions. The benefits of improved pumping efficiency include reduced reliance on both the electricity grid and renewable energy supplies and improved operational reliability. Furthermore, saving a single unit of pumping energy can actually save more than ten times that energy in fuel. Due to the inefficiencies of a mostly centralized electricity transmission system, 100 units of fuel input at the power station are required to achieve 9.5 units of energy output at the pumping system.[2] But the reverse is also true: saving 9.5 units of energy output at the pump could save 100 units of energy at the power station.[3]

Generally, smaller pumping systems tend to be more inefficient than large ones. Small pumping systems typically make up only a small fraction of the total cost of an industrial operation and thus receive relatively little design attention. However, the significance of small pumping systems cannot be overlooked. There are many more small and medium-sized enterprises than there are large enterprises. Thus it is likely that there are a lot more small pumping systems than large pumping systems, especially since small enterprises almost exclusively use small pumping systems and large enterprises use both small and large pumping systems. Large pumping systems, with power ratings in the order of kilowatts and megawatts, that are poorly designed and managed can attract very high and unnecessary costs. Consequently, large pumping system design is typically quite disciplined, with more attention paid to factors such as minimum velocities, thermal expansion, pipe work and maintenance. Still, there are very few pumping systems that wouldn't benefit from Whole System Design (WSD).

Worked example overview

Pumping systems are a subgroup of motor systems. Other subgroups include ventilation systems (fans), compressed air systems (compressors) and conveyor systems (gears, pulleys and belts). Further information about the Whole System Design of motor systems is available in The Natural Edge Project's freely available online textbook *Energy Transformed: Sustainable Energy Solutions for Climate Change Mitigation*, 'Lecture 3.1: Opportunities for improving the efficiency of motor systems'.[4]

The following worked example focuses on pumping systems. Specifically, it provides a worked mathematical example similar to a well-known WSD case study, 'Pipes and pumps', which is briefly described in the following extract from *Natural Capitalism*:[5]

In 1997, leading American carpet maker, Interface Inc, was building a factory in Shanghai. One of its industrial processes required 14 pumps. In optimizing the design, the top Western specialist firm sized those pumps to total 95 horsepower. But a fresh look by Interface/Holland's engineer Jan Schilham, applying methods learned from Singaporean efficiency expert Eng Lock Lee, cut the design's pumping power to only 7 horsepower – a 92 per cent or 12-fold energy saving – while reducing its capital cost and improving its performance in every respect.

The new specifications required two changes in design. First, Schilham chose to deploy big pipes and small pumps instead of the original design's small pipes and big pumps. Friction falls as nearly the fifth power of pipe diameter, so making the pipes 50 per cent fatter reduces their friction by 86 per cent. The system then needs less pumping energy – and smaller pumps and motors to push against the friction. If the solution is this easy, why weren't the pipes originally specified to be big enough? Because of a small but important blind spot: traditional optimization compares the cost of the fatter pipe only with the value of the saved pumping energy. This comparison ignores the size, and hence the capital cost, of the equipment – pump, motor, motor-drive circuits and electrical supply components – needed to combat the pipe friction. Schilham found he needn't calculate how quickly the savings could repay the extra up-front cost of the fatter pipe, because capital cost would fall more for the pumping and drive equipment than it would rise for the pipe, making the efficient system as a whole cheaper to construct.

Second, Schilham laid out the pipes first and then installed the equipment, in reverse to how pumping systems are conventionally installed. Normally, equipment is put in some convenient and arbitrary spot, and the pipe fitter is then instructed to connect point A to point B. The pipe often has to go through all sorts of twists and turns to hook up equipment that's too far apart, turned the wrong way, mounted at the wrong height or separated by other devices installed in between. The extra bends and the extra length make friction in the system about three to six times higher than it should be. The pipe fitters don't mind the extra work: they're paid by the hour, they mark up the pipe and fittings, and they won't have to pay the pumps' capital or operating costs.

By laying out the pipes before placing the equipment that the pipes connect, Schilham was able to make the pipes short and straight rather than long and crooked. That enabled him to exploit their lower friction by making the pumps, motors, inverters and electricals even smaller and cheaper.

The fatter pipes and cleaner layout yielded not only 92 per cent lower pumping energy at a lower total capital cost, but also simpler and faster construction, less use of floor space, more reliable operation, easier maintenance, and better performance. As an added bonus, easier thermal insulation of the straighter pipes saved an additional 70 kilowatts of heat loss, enough to avoid burning about a pound of coal every two minutes, with a three-month payback.

Schilham marvelled at how he and his colleagues could have overlooked such simple opportunities for decades. His redesign required, as inventor Edwin Land used to say, 'not so much having a new idea as stopping having an old idea'. The old idea was to 'optimize' only part of the system – the pipes – against only one parameter – pumping energy. Schilham, in contrast, optimized the whole system for multiple benefits – pumping energy expended plus capital cost saved. (He didn't bother to value explicitly the indirect benefits mentioned, but he could have.)

Figure 6.1 shows the setting for the worked example, a typical production plant scenario where a pumping system would be used. In Figure 6.1, a known fluid at temperature T must be moved from point 1 in reservoir A to point 2 at the tap with a target exit volumetric flow rate of Q. Between the reservoir and tap is a window (fixed into the wall) and a machine press (moveable).

Recall the elements of applying a WSD approach discussed in Chapters 4 and 5:

1 Ask the right questions.
2 Benchmark against the optimal system.
3 Design and optimize the whole system.
4 Account for all measurable impacts.
5 Design and optimize subsystems in the right sequence.
6 Design and optimize subsystems to achieve compounding resource savings.
7 Review the system for potential improvements.

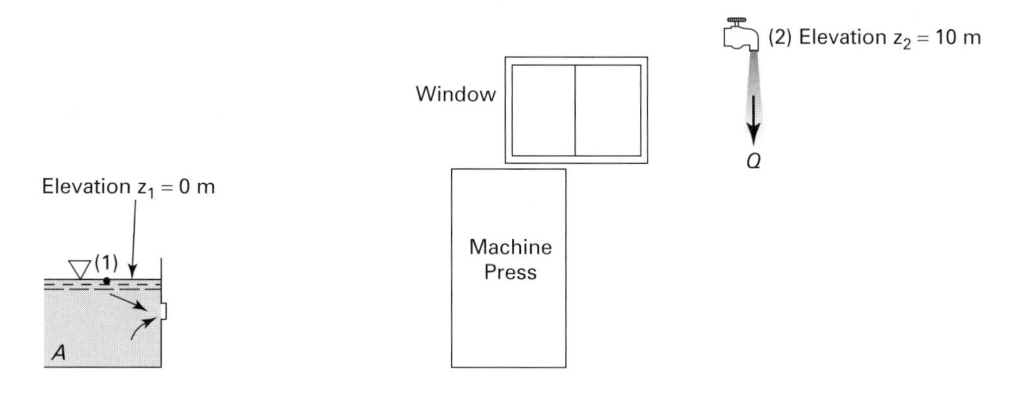

Source: Adapted from Munson, Young and Okiishi (1998), pp512 and 522

Figure 6.1 *A typical production plant scenario*

Design challenge

Consider water at 20°C flowing from reservoir A, through the system in Figure 6.1, to a tap with a target exit volumetric flow rate of Q = 0.001 m^3/s. Select suitable pipes based on pipe diameter, D, and a suitable pump based on pump power, P, and calculate the cost of the system.

Design process

The following sections of this chapter present:

1 *General solution*: A solution for any single-pump, single-pipe system with the given constraints;
2 *Conventional design solution*: Conventional system with limited application of the elements of WSD;
3 *WSD solution*: Improved system using the elements of WSD;
4 *Performance comparison*: Comparison of the economic and environmental costs and benefits.

8 Model the system.
9 Track technology innovation.
10 Design to create future options.

The following worked example will demonstrate how the elements can be applied to pumping systems using two contrasting examples: a conventional pumping versus a WSD pumping system. The application of an element will be indicated with a shaded box.

General solution

The calculations in this chapter use several variables, as defined in Table 6.1.

Figure 6.2 shows a typical single-pump, single-pipe solution, which includes the following features:

- The system accommodates the pre-existing floor plan (window) and equipment (machine press) in the plant.
- Reservoir A exit is very well rounded.
- The diameter of every pipe is D.
- A globe valve, which acts as an emergency cut-off and stops the flow for maintenance purposes, is fully open during operation.
- The existing tap is replaced by a tap with an exit diameter of D.

The Natural Edge Project provides a freely available online 'Appendix A' document,[6] containing equations, figures and tables that are applied to the General, Conventional and WSD solutions in the following sections. The majority of these equations, figures and tables are taken from Munson, Young and Okiishi (1998)[7] and can also be applied to similar pipe and pump systems.

Table 6.1 *Symbol nomenclature*

Symbol	Description	Unit	Symbol	Description	Unit
p	Pressure	Pa	L	Pipe length	m
ϱ	Density	kg/m^3	D	Pipe diameter	m
g	Acceleration due to gravity	$9.81 m/s^2$	Re	Reynolds number	
α	Kinetic energy coefficient		μ	Dynamic viscosity	Ns/m^2
V	Average velocity	m/s	ε	Equivalent roughness	mm
z	Height	m	K_L	Loss coefficient	
h	Head loss	m	A	Pipe cross sectional area	m^2
f	Friction factor		P	Power	W

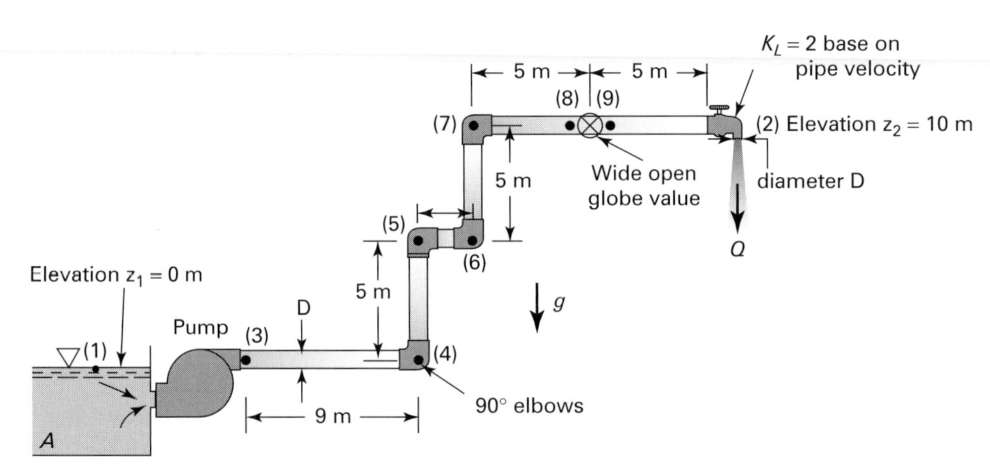

Figure 6.2 *A typical single-pump, single-pipe configuration*

Element 8: Model the system

The energy balance between point 1 and point 2 in the system is given by Bernoulli's Equation:[8]

$$p_1/\varrho g + \alpha_1 V_1^2/2g + z_1 + \Sigma P_i/\varrho g A_i V_i = p_2/\varrho g + \alpha_2 V_2^2/2g + z_2 + \Sigma f_i (L_i/D_i)(V_i^2/2g) + \Sigma K_{Li} V_i^2/2g$$

Some simplifications and substitutions can be made based on the configuration of the system:

$p_1 = p_2 = 0$ (atmospheric pressure)

$V_1 = 0$

$z_1 = 0$

Since reservoir A exit is very well rounded, assume the corresponding component loss is negligible

Since the diameter of every pipe is D (constant):[9]

- The cross-sectional area of every pipe is A

- The average velocity of the fluid downstream of the pump is constant and equal to V_2

The pipes are considered to be a single pipe of length L

Assume the pipe is completely full of water since there is no downward flow[10]

Assume that pipes are available in the lengths indicated in Figure 6.2

Assume that head losses through pump connectors, tap connectors and reservoir A exit are negligible

Thus, the energy balance reduces to:

$$P/pgAV_2 = \alpha_2 V_2^2/2g + z_2 + f\,(L/D)(V_2^2/2g) + V_2^2/2g\,(\Sigma K_{Li})$$

The design variables to be determined are:

Pump power, P

Pipe diameter, D

The known variables are:

ϱ physical property of water

z_2 from system plan

L from system plan

K_{Li} function of pipe/reservoir interface geometry and component geometry

V_2 can be eliminated from the energy balance equation by substituting for functions of Q and D using:

$$V_2 = Q/A$$

and

$$A = \Pi D^2/4$$

Substituting and making pump power, P, the subject of the equation gives:

$$P = (8\varrho Q^3/\Pi^2 D^4)\,[\alpha_2 + f\,(L/D) + \Sigma K_{Li}] + \varrho g Q z_2$$

The friction factor, f, is dependent on the Reynolds number, Re:

$$Re = \varrho V_2 D/\mu$$

Substituting for V_2 gives:

$$Re = 4\varrho Q/\Pi D\mu$$

Where μ is a known physical property of water. For a turbulent flow ($Re > 4000$), the equivalent roughness of the interior of the pipe, ε, a known physical property of the pipe, is required to determine f.

We now have the relationship between pump power, P, and pipe diameter, D, in terms of known variables for the system in Figure 6.2.

Conventional design solution

Select suitable pipes and pumps for the system

For water at 20°C:[11]

$$\varrho = 998.2\text{kg/m}^3$$

$$\mu = 1.002 \times 10^{-3}\text{Ns/m}^2$$

Calculating Reynolds number:

$$Re = 4(998.2\text{kg/m}^3)(0.001\ \text{m}^3/\text{s})/\ \Pi D(1.002 \times 10^{-3}\text{Ns/m}^2)$$

$$Re = 1268/D$$

The flow is turbulent ($Re > 4000$) for $D < 0.317$m. A pipe of diameter $D = 0.317$m is much larger than what is suitable for the system[12] in Figure 6.2; thus it is safe to assume that the flow is turbulent. Since turbulent velocity profiles are nearly uniform across the pipes, we assume $\alpha_1 = \alpha_2 = 1$.

For 90°C threaded elbows:[13]

$$K_{L4} = K_{L5} = K_{L6} = K_{L7} = 1.5$$

For a fully open globe valve:[14]

$K_{LV} = 10$

For the tap:

$K_{LT} = 2$

The energy balance equation becomes:

$$P = [8(998.2kg/m^3)(0.001m^3/s)^3/\pi^2D^4][1 + f(30/D) + (1.5 \times 4 + 10 + 2)] + (998.2kg/m^3)(9.81\ m/s^2)(0.001\ m^3/s)(10m)$$

$$P = (8.0911 \times 10^{-7}/D^4)[f(30/D) + 19] + 97.923$$

Suppose drawn copper tubing of diameter D = 0.015m was selected for the pipes. Substituting into the Reynolds number equation gives:

$$Re = 1268.411/(0.015m) = 84561$$

For drawn tubing:[15]

$$\varepsilon = 0.0015mm$$

Thus:

$$\varepsilon/D = 0.0015/15 = 0.0001$$

Using the Moody chart,[16] Re = 84561 and $\varepsilon/D = 0.0001$ give:

$$f = 0.0195$$

Substituting D = 0.015m and f = 0.0195 into the equation for pump power gives:

$$P = (8.0911 \times 10^{-7}/(0.015m)^4)[0.0195(30/(0.015m)) + 19] + 97.923 = 1025\ W$$

That is, for the system in Figure 6.2, if drawn copper tubing of diameter D = 0.015m is used for the pipes, then a pump of power P = 1025 W is required to generate an exit volumetric flow rate of $Q = 0.001m^3/s$.

We can select pump model:[17]

Waterco Hydrostorm Plus 150[18] at P = 1119 W (1.5 hp)

We can select pipe:[19]

Hard drawn copper tube (6M length) T24937 at D = 15mm (5/8 in)

Calculate the cost of the system

Copper pipe T24937 costs AU$57.12 per 6m.[20] Therefore the cost of 30m of copper pipe is:

Pipe cost = (AU$57.12 per 6m)(30m)/6 = AU$285.60

Standard radius 90° elbows of 15mm (5/8 in) diameter J00231 cost of AU$2.34 each.[21] Therefore the total cost of the elbows is:

Elbow cost = (AU$2.34)(4) = AU$9.36

A globe valve of diameter 15mm (5/8 in):[22]

Estimated globe valve cost = AU$13 (US$10)

A tap of exit diameter 0.015m:[23]

Tap cost = AU$6.70

Installation costs for 8hrs at AU$65/hr gives:

Installation costs = (AU$65/hr)(8 hrs) = AU$520

The Waterco Hydrostorm Plus 150:[24]

Pump cost = AU$616

Thus, the total capital cost of the system is:

Capital cost = AU$285.60 + AU$9.36 + AU$13+ AU$6.70 + AU$520 + AU$546 = AU$1451

To calculate running costs for the selected electrically powered pump, the following values are used:

Pump efficiency for an electrical pump: 47%[25]

Cost of electricity: AU$0.1/kWh (2006 price for large energy users)

For the Waterco Hydrostorm Plus 150 pump running at output power P = 1025 W, the monthly pump running costs for 12 hrs/day, 26 days/month are:

Running cost = (AU$0.1/kWh)(1.025kW) (12 hrs/day)(26 day/mth)/(0.47) = AU$68/mth

WSD Solution

> *Element 1: Ask the right questions*

Is the conventional solution optimal for the whole system? What are the factors of the whole system that need to be considered? The conventional design solution was suboptimal for two reasons:

1 The pipe configuration introduced head losses that could be avoided; and
2 The selection procedure for pipe diameter, D, and pump power, P, did not address the *whole system*.

Redesign the pipes and pump system with less head loss

> *Element 7: Review the system for potential improvements*

Items to consider:

- From Bernoulli's equation, $Power \propto \frac{1}{D^4}$, so increasing diameter dramatically reduces power required.
- Can the system be designed with less bends?
- Can the system be designed with more-shallow bends?
- Is it worthwhile moving the plant equipment (machine press)?
- Is an alternative pipe material more suitable?
- Is there a more suitable valve? Do we even need a valve?

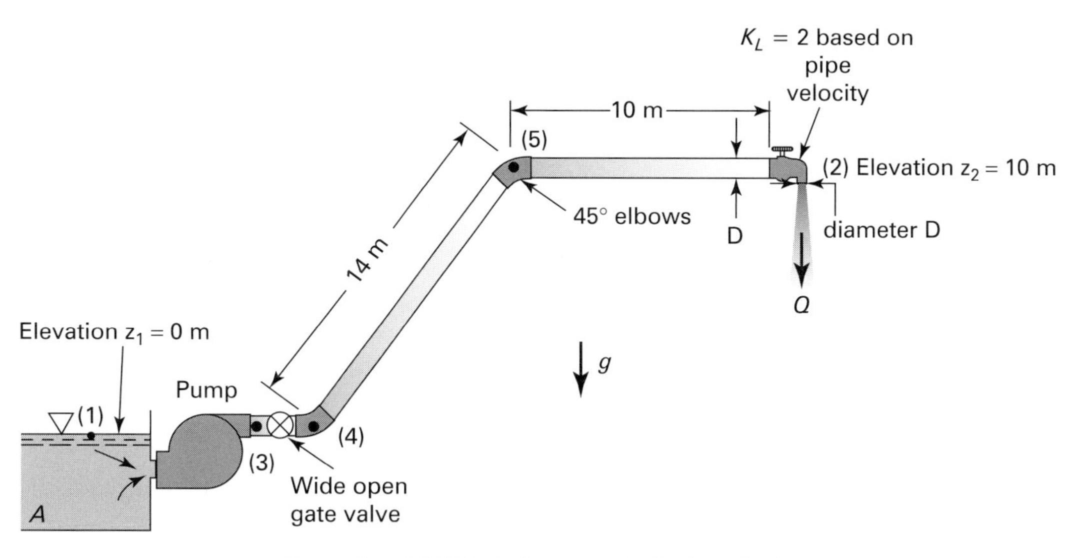

Figure 6.3 *A WSD single-pump, single-pipe solution*

Select suitable pipes and pumps for the system

Since the conditions at point 1 and point 2 in Figure 6.3 are the same as in Figure 6.2, and a single pump and single pipe are used, the energy balance equation for the general solution is applicable:

$$P = (8pQ^3/\Pi^2D^4)\ [\alpha_2 + f\ (L/D) + \Sigma\ K_{Li}] + pgQz_2$$

For 45° threaded elbows:[26]

$$K_{L4} = K_{L5} = 0.4$$

For a fully open gate valve:[27]

$$K_{LV} = 0.15$$

For the tap:

$$K_{LT} = 2$$

The energy balance equation becomes:

$$P = [8(998.2kg/m^3)(0.001m^3/s)^3/\Pi^2D^4]\ [1 + f\ (24/D) + (0.4 \times 2 + 0.15 + 2)] + (998.2kg/m^3)(9.81m/s^2)(0.001m^3/s)(10m)$$

$$P = (8.0911 \times 10^{-7}/D^4)[f\ (24/D) + 3.95] + 97.923$$

Suppose, instead, a drawn copper pipe of diameter D = 0.03m (double the diameter in the conventional solution) was selected. Substituting into the Reynolds number equation gives:

$$Re = 1268.411/(0.03m) = 42280$$

For drawn tubing:[28]

$$\varepsilon = 0.0015mm$$

Thus:

$$\varepsilon/D = 0.0015/30 = 0.00005$$

Using the Moody chart,[29] Re = 42280 and ε/D = 0.00005 give:

$$f = 0.0215$$

Substituting D = 0.03m and f = 0.0215 into the equation for pump power gives:

$$P = (8.0911 \times 10^{-7}/(0.03m)^4)[0.0215\ (24/(0.03m)) + 3.95] + 97.923 = 119\ W$$

That is, for the system in Figure 6.3, if drawn copper tubing of diameter D = 0.03m is used for the pipes, then a pump of power P = 119 W is required to generate an exit volumetric flow rate of Q = 0.001m³/s.

We can select pump model:[30]

Monarch ESPA Whisper 500[31] at
P = 370 W (0.5 hp)

We can select pipe:[32]

Hard drawn copper tube (6M length)
T22039 at D = 31.75mm (1¼ in)

Is this the optimal solution for the whole system?

Consider the effect of other pipe diameters and pump powers

Element 3: Design and optimize the whole system

Other combinations of pipe diameter and pump power[33] that suit the system can be selected in a similar way, as in Table 6.2:

Table 6.2 *Pump power calculated for a spectrum of pipe diameters*

D (m)	Re	ε/D	F	P (W)
0.015	84561	0.0001	0.0195	660
0.02	63421	0.000075	0.0205	242
0.025	50736	0.00006	0.0210	148
0.03	42280	0.00005	0.0215	119
0.04	31710	0.0000375	0.0230	104

Calculate the cost of the system

The capital and running costs for each pipe and pump combination are shown in Table 6.3. The costs are calculated in a similar way as for the conventional solution. The efficiency of the Monarch ESPA Whisper 1000 is approximated at 42 per cent[34] and the efficiency of the Monarch ESPA Whisper 500 is approximated at 40 per cent.[35] The life-cycle economic cost of each solution is estimated as the net present value (NPV) calculated over a life of 50 years and at a discount rate of 6 per cent.

Table 6.3 shows that the solution with D = 0.015m has the lowest capital cost by a relatively small margin, but the highest life-cycle cost by a factor of 2–3. Given the estimation errors in our calculations, the life-cycle cost for the solution with D = 0.03m is about the same as that for a system with D = 0.04m. However, the capital cost is about AU$200 less and would therefore incur smaller economic stress up front.[36] Hence, for the optimal pipe and pump combination for the system in Figure 6.3 we can select:

ESPA Whisper 500 pump at P = 370 W (0.5 hp)

T22039 hard drawn copper pipe at
D = 31.75mm (1¼ in)

Summary: performance comparisons

Pump power and cost

A side-by-side comparison of the conventional design solution system and the WSD solution in Table 6.4 highlights the substantially different results that each approach achieves.

The life-cycle cost of the WSD solution is about five times smaller than for the conventional solution. Since the capital costs of both solutions are similar, it is obvious that the cost savings for the WSD solution arise from the lower required pumping power and hence running cost. This example demonstrates the dominance of running costs over capital costs – a relationship that is common for many resource-consuming systems. The power reduction was made possible by the inclusion of two additional steps in the design and selection process:

Step 1: Redesign the pipes and pump system with less head loss; and
Step 2: Consider the effect of other pipe diameters and pump powers.

Table 6.3 *Summary of system costs for a range of pump types[37] and pipe diameters[38]*

D (m)	Pipes and components cost	P (W)	Pump selected	Pump cost	Total capital cost	Running cost	Life cycle cost (–NPV[39])
0.015	$602	660	Monarch ESPA Whisper 1000	$357	$959	$49/mth	$10,821
0.02	$745	242	Monarch ESPA Whisper 500	$331	$1076	$19/mth	$4873
0.025	$827	148	Monarch ESPA Whisper 500	$331	$1158	$12/mth	$3480
0.03	$914	119	Monarch ESPA Whisper 500	$331	$1245	$9/mth	$3112
0.04	$1126	104	Monarch ESPA Whisper 500	$331	$1457	$8/mth	$3089

Table 6.4 *Comparing the costs of the two solutions*

Solution	D (m)	Pipes and components cost	P (W)	Pump cost	Total capital cost	Running cost	Life cycle cost (–NPV)
Conventional	0.015	$835	1025	$616	$1451	$61/mth	$15,129
WSD	0.03	$914	119	$331	$1245	$9/mth	$3112

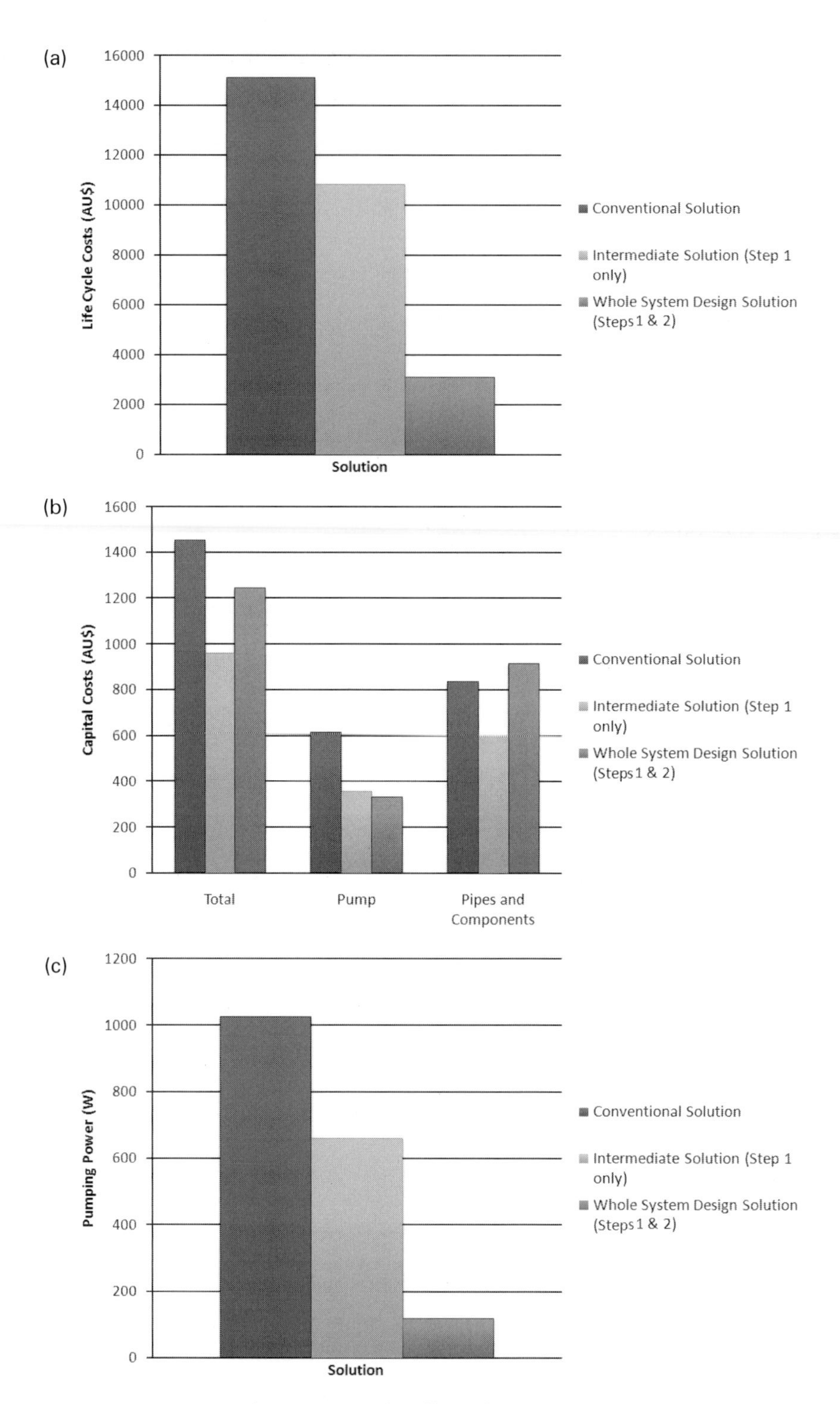

Figure 6.4 *Comparing the effects of Step 1 and Step 2*

Step 1 optimized the system configuration and yielded system wide improvement, regardless of the pipe diameter selected. Even with the same pipe diameter as the conventional solution (D = 0.015m), the WSD solution has a 28 per cent lower pipes and components cost, requires 36 per cent less power, has a 34 per cent lower capital cost, and comes in about 28 per cent cheaper over its life, as shown in Figure 6.4.

Step 2 optimized the pipe diameter and pump selection process. Notably, the larger diameter pipes reduced the total required pumping power of the system. The second step resulted in a further 82 per cent reduction in power and 71 per cent reduction in life-cycle cost, as shown in Figure 6.4.

In total, the WSD solution uses 88 per cent less power, costs 79 per cent less over its life, and is cheaper to purchase and install than the conventional solution.

Multiple benefits

Element 4: Account for all measurable impacts

A number of other benefits arise from designing the pumping system such that it is 'short, fat and straight' rather than 'long, thin and bent':

- *More floor space is available*: Less piping covering the floors of industrial sites means more space is available to work in, as well as improving the safety of the work environment.
- *More reliable operation*: Less bends and valves in piping reduces the likelihood of parts failing. Reducing friction in the piping means that less energy is lost to adding physical stress to the piping system, thereby increasing the life of the system. Since less power is required, the motor driving the pump doesn't need to work as hard.
- *Easier maintenance*: With short and straight pipes, maintenance workers can get into the system with relative ease, as opposed to negotiating a maze of piping in the conventional solution.
- *Better performance*: A much greater percentage of energy used in the system is converted into useful work. A system that is more reliable and easy to maintain provides consistently high performance relative to conventional systems.

Factors to consider for larger systems

Extra considerations for larger pipes

The pipe sizes considered in this worked example can be installed and mounted without restriction. However, a few notes should be made about larger pipes:

- A structure design permit may be required before mounting the pipe to an existing structure. Attaining the permit may incur a cost.
- Large pipes are heavier and thus may require additional mounting support, which may incur a cost.
- Pipes larger than about 0.05m (2 in) in diameter may require stress analysis to account for the effects of thermal expansion. Tables that suggest when stress analysis should be performed are available. The tables usually consider pipe diameter and fluid temperature.
- Long straight pipes experience more wall stress than shorter, bent pipes. Systems with long straight pipes can also result in higher forces and resulting moments on inertia on the fixed nozzles of equipment, especially when the endpoints of the system are under pressure (say in a tank as opposed to open air). In these cases, expansion joints and bellowed nozzles can be incorporated to advantage, with the key consideration being to make bends as smooth as possible.

Site planning

In this worked example, it was assumed that reservoir A, the pump and the tap were to remain where there were. Sometimes, the location of such features is arbitrary, as in the 'Pipes and pumps' example in *Natural Capitalism*,[40] so their location can be governed by the piping system. In other cases, however, other factors can influence where these features as well as the pipes should be located. For example, a pipe and pump system can share many resources with other equipment and systems. These resources include shelter, electrical cable route, drainage systems and access ways for maintenance. Accounting for these factors is an example of Element 3, 'Design and optimize the whole system', and Element 4, 'Account for all measurable impacts'.

Larger cost reductions

In this worked example, the small amount of required pumping power[41] did not lend itself to a good demonstration of pump capital cost savings. Table 6.3 shows that even though required pumping power fell by a factor of more than two between the solution where D = 0.02m and the solution where D = 0.04m, the same pump was used for all solutions and hence the pump capital cost was the same. In a larger system, the required pumping power falls over a larger range, for which there is a variety of pumps that can be selected.

Systems powered by internal combustion engines

Some moderate sized systems use pumps powered by internal combustion engines (ICEs). The size of ICE pumps start at about 1.5kW output power. They are usually cheaper to purchase but more expensive to run than the equivalent electric pump. Consequently, moderate sized ICE pump systems have even greater potential for cost savings.

To demonstrate, consider a conventional, ICE-powered system that requires 10kW of pumping power. We have shown that WSD can reduce the pumping power of a conventional system by 88 per cent. The 10kW conventional system can therefore be redesigned as a 1.2kW system, which means the 10kW ICE pump costing about AU$12,700[42] can be replaced with a 1.5kW electric pump costing AU$616.[43]

Now, since the required pumping power is reduced by 88 per cent, the running costs are then reduced by 88 per cent. Furthermore, an *additional* saving arises since the electrical pump is at least twice as efficient as the ICE pump (20–26 per cent[44]), while the cost per unit energy is about the same for electricity (AU$0.10/kWh for large energy users, AU$0.17 for domestic users) and petrol (AU$0.14/kWh at AU$1.30 per litre).

Effectively, the lower power consumption of the WSD solution makes solutions viable that bring with them additional benefits and that are otherwise too expensive.

To calculate the cost per unit energy of petrol, the following values are used:

Energy value of petrol: 34 MJ/litre[45]

Cost of petrol: AU$1.30/litre (2006 price at the pump)

The cost per unit energy for petrol is:

Cost per unit energy = (AU$1.30/litre)/(34,000,000 J/litre) = 3.8235×10^{-8} AU$/J

Converting to units of AU$/kWh:

Cost per unit energy = [(3.8235×10^{-8} AU$/J)/(1 s)](1000 W/kW)(3600 s/hr) = AU$0.14/kWh.

Notes

1. Hawken, P., Lovins, A. B. and Lovins, L. H. (1999) *Natural Capitalism: Creating the Next Industrial Revolution*, Earthscan, London, p115.
2. Hawken, P., Lovins, A. B. and Lovins, L. H. (1999) *Natural Capitalism: Creating the Next Industrial Revolution*, Earthscan, London, p121.
3. Hawken, P., Lovins, A. B. and Lovins, L. H. (1999) *Natural Capitalism: Creating the Next Industrial Revolution*, Earthscan, London, p121.
4. Smith, M., Hargroves, K., Stasinopoulos, P., Stephens, R., Desha, C. and Hargroves, S. (2007) *Energy Transformed: Sustainable Energy Solutions for Climate Change Mitigation*, The Natural Edge Project, Australia, 'Lecture 3.1: Opportunities for improving the efficiency of motor systems', www.naturaledgeproject.net/Sustainable_Energy_Solutions_Portfolio.aspx, accessed 10 April 2008.
5. Hawken, P., Lovins, A. B. and Lovins, L. H. (1999) *Natural Capitalism: Creating the Next Industrial Revolution*, Earthscan, London, p121.
6. Stasinopoulos, P., Smith, M., Hargroves, K. and Desha, C. (2007) *Whole System Design – An Integrated Approach to Sustainable Engineering*, The Natural Edge Project, Australia, 'Unit 6: Worked Example 1 – Industrial Pumping Systems', Appendix A, www.naturaledgeproject.net/Whole_System_Design.aspx, accessed 10 April 2008.
7. Munson, B. R., Young, D. F. and Okiishi, T. H. (1998) *Fundamentals of Fluid Mechanics* (third edition), Wiley and Sons, New York.
8. This equation represents the four kinds of energy changes associated with fluid flow through a pipe and pump system: 1) Pressure, kinetic energy and potential energy changes, 2) Friction losses, 3) Component losses, and 4) Pumping gains. Alternatively, see The Natural Edge Project's online Appendix A, www.naturaledge

project.net/Whole_System_Design.aspx, accessed 10 April 2008.

9 A and V are dependent on D.

10 This assumption aims to omit two possible situations where air is present in the pipe. The first situation occurs when the portion of the pipe nearest the tap contains air because there isn't enough water to fill the pipe. In practice, this situation can be overcome by turning off the tap before turning off the pump when shutting down. The second situation occurs when water and air share space in the pump at the same point (but don't mix). This configuration is often referred to as a 'channel' configuration because of the resemblance to a channelled waterway such as a river (open channel) or a sewage pipe (closed channel). Since water is denser than air, water will occupy the bottom side of the channel and air will occupy the top side; and since all flow is either horizontal or against gravity then, given enough water and an outlet for the air to escape (tap), the pipe will likely be filled with water.

11 Munson, B. R., Young, D. F. and Okiishi, T. H. (1998) *Fundamentals of Fluid Mechanics* (third edition), Wiley and Sons, New York, p853. Alternatively, see The Natural Edge Project's online Appendix A, www.naturaledgeproject.net/Whole_System_Design.aspx, accessed 10 April 2008.

12 The pipe diameter for the system in Figure 6.2 is likely to be no less than D = 0.01m and no more than D = 0.05m.

13 Munson, B. R., Young, D. F. and Okiishi, T. H. (1998) *Fundamentals of Fluid Mechanics* (third edition), Wiley and Sons, New York, p505. Alternatively, see The Natural Edge Project's online Appendix A, www.naturaledgeproject.net/Whole_System_Design.aspx, accessed 10 April 2008.

14 Munson, B. R., Young, D. F. and Okiishi, T. H. (1998) *Fundamentals of Fluid Mechanics* (third edition), Wiley and Sons, New York, p505. Alternatively, see The Natural Edge Project's online Appendix A, www.naturaledgeproject.net/Whole_System_Design.aspx, accessed 10 April 2008.

15 Munson, B. R., Young, D. F. and Okiishi, T. H. (1998) *Fundamentals of Fluid Mechanics* (third edition), Wiley and Sons, New York, p492. Alternatively, see The Natural Edge Project's online Appendix A, www.naturaledgeproject.net/Whole_System_Design.aspx, accessed 10 April 2008.

16 Munson, B. R., Young, D. F. and Okiishi, T. H. (1998) *Fundamentals of Fluid Mechanics* (third edition), Wiley and Sons, New York, p493. Alternatively, see The Natural Edge Project's online Appendix A, www.naturaledgeproject.net/Whole_System_Design.aspx, accessed 10 April 2008.

17 See PumpShop website at www.pumpshop.com.au/, accessed 11 August 2005.

18 Waterco (2004) *Hydrostorm Plus Pool and Spa Pumps*, Waterco, p2, www.waterco.com.au/CMS/uploads/brochures/HydroPumpsZZB1285.pdf, accessed 11 April 2008. Waterco shows that this pump provides near maximum head at Q = 60l/min ($0.001m^3/s$), so this pump will satisfy both head and flow rate requirements.

19 Kirby (2004) *Copper Tube and Fittings*, Kirby, Australia, http://www.kirbyrefrig.com.au/pdf/200401/9coppert.pdf, accessed 10 August 2008.

20 Kirby (2004) *Copper Tube and Fittings*, Kirby, Australia, http://www.kirbyrefrig.com.au/pdf/200401/9coppert.pdf, accessed 10 August 2008.

21 Kirby (2004) *Copper Tube and Fittings*, Kirby, Australia, http://www.kirbyrefrig.com.au/pdf/200401/9coppert.pdf, accessed 10 August 2008.

22 Interpolated from available 'globe valve' sizes supplied by A. Y. McDonald MFG. Co. at www.aymcdonald.com, accessed 11 August 2005.

23 'Tap brass (hose cock)' supplied by Wet Earth at www.wetearth.com.au, accessed 11 August 2005.

24 See PumpShop website at www.pumpshop.com.au/, accessed 11 August 2005.

25 ESPA (2000) 'SILENT Series; TYPHOON Series: Swimming pool pumps: Instruction manual', Monarch Pool Systems, p2, www.monarchpoolsystems.com/manuals/PDF/Espa-manual.pdf, accessed 11 July 2006. This value is an approximation based on the data given by ESPA for the Silent 75M.

26 Munson, B. R., Young, D. F. and Okiishi, T. H. (1998) *Fundamentals of Fluid Mechanics* (third edition), Wiley and Sons, New York, p505. Alternatively, see The Natural Edge Project's online Appendix A, www.naturaledgeproject.net/Whole_System_Design.aspx, accessed 10 April 2008.

27 Munson, B. R., Young, D. F. and Okiishi, T. H. (1998) *Fundamentals of Fluid Mechanics* (third edition), Wiley and Sons, New York, p505. Alternatively, see The Natural Edge Project's online Appendix A, www.naturaledgeproject.net/Whole_System_Design.aspx, accessed 10 April 2008.

28 Munson, B. R., Young, D. F. and Okiishi, T. H. (1998) *Fundamentals of Fluid Mechanics* (third edition), Wiley and Sons, New York, p492. Alternatively, see The Natural Edge Project's online Appendix A document. Pipes of diameter 0.01–0.04m are available in a few different materials, including copper, steel and aluminium. Munson, Young and Okiishi suggest that drawn metal tubing, such as the copper pipes incorporated in the conventional solution, is the smoothest of the suitable pipes for the Design Challenge. Although plastic pipes are the smoothest ($\varepsilon \approx 0$), they are

also generally larger than what is required, starting at diameters of about 0.05m (1 in).

29 Munson, B. R., Young, D. F. and Okiishi, T. H. (1998) *Fundamentals of Fluid Mechanics* (third edition), Wiley and Sons, New York, p493. Alternatively, see The Natural Edge Project's online Appendix A, www.naturaledgeproject.net/Whole_System_Design.aspx, accessed 10 April 2008.

30 See PumpShop website at www.pumpshop.com.au/, accessed 11 August 2005.

31 Monarch Pool Systems (n.d.) 'Whisper Series: Swimming pool pumps', Monarch Pool Systems, p2, www.monarchpoolsystems.com/products/Low%20Res%20PDFs/Whisper.pdf, accessed 11 April 2008. Monarch Pool Systems show that this pump provides near-maximum head at Q = 60l/min (0.001m^3/s), so this pump will satisfy both head and flow rate requirements.

32 Kirby (2004) *Copper Tube and Fittings*, Kirby, Australia, http://www.kirbyrefrig.com.au/pdf/200401/9coppert.pdf, accessed 10 August 2008.

33 Only systems with diameter up to D = 0.04m are shown. At higher diameters the power savings become small. For example D = 0.05m gives P = 102 W; and D = 0.06m gives 100 W.

34 ESPA (2000) 'SILENT Series; TYPHOON Series: Swimming pool pumps: Instruction manual', Monarch Pool Systems, p2, www.monarchpoolsystems.com/manuals/PDF/Espa-manual.pdf, accessed 11 July 2006. This value is an approximation based on the data given by ESPA for the Silent 30M.

35 ESPA (2000) 'SILENT Series; TYPHOON Series: Swimming pool pumps: Instruction manual', Monarch Pool Systems, p2, www.monarchpoolsystems.com/manuals/PDF/Espa-manual.pdf, accessed 11 July 2006. This value is an approximation based on the data given by ESPA, which shows a trend of decreasing efficiency with decreasing power capacity.
Alternatively, the economic uncertainty associated with spreading the system cost over a 50-year period may be a greater stress than having to pay more upfront. Consequently, in this worked example, either solution is as good as the other.

36 See PumpShop website at www.pumpshop.com.au/, accessed 11 August 2005.

37 Kirby (2004) *Copper Tube and Fittings*, Kirby, Australia, http://www.kirbyrefrig.com.au/pdf/200401/9coppert.pdf, accessed 10 August 2008.

38 Negative values for NPV are actually costs.

39 Hawken, P., Lovins, A. B. and Lovins, L. H. (1999) *Natural Capitalism: Creating the Next Industrial Revolution*, Earthscan, London.

40 The optimized WSD solution required the same amount of power (119 W) as a bright incandescent lamp.

41 '10kW (13 hp) Fire 02.5F13K2V pump' supplied by A. Y. McDonald MFG. Co. at www.aymcdonald.com, accessed 11 August 2005.

42 '1.5kW (2 hp) Waterco Hydrostorm Plus 200 pump' supplied by PumpShop at www.pumpshop.com.au/, accessed 11 August 2005.

43 Evans, R., Sneed, R. E. and Hunt, J. H. (1996) *Pumping plant performance evaluation*, North Carolina Cooperative Extension Service, www.bae.ncsu.edu/programs/extension/evans/ag452-6.html, accessed 27 June 2006. This value is an overestimate. The data are for an internal combustion engine only, and do not include any mechanical losses associated with the coupling of the engine to the pump or the pump itself.

44 Moorland School (n.d.) *Petrol*, Moorland School, www.moorlandschool.co.uk/earth/petrol.htm, accessed 27 June 2006.

7

Worked Example 2 – Passenger Vehicles

Significance of the Automotive Industry and Vehicle Design

Few industries in the manufacturing sector are under more pressure at present than the automotive industry. Companies are not only having to remain competitive in what many consider an over-mature market, but are also pushed by government legislation and consumer demand for vehicles that pollute less and use fuel alternatives.

Australia's appetite for petrol-fuelled, family-sized sedans is a large contributor to the nation's greenhouse emissions. In 2002 the Australian transport sector was responsible for 79 million tons of greenhouse gas emissions, comprising 13 per cent of Australia's total emissions. Around 88 per cent of these emissions came from road transport vehicles such as cars, buses and trucks.[1] This figure is expected to increase by 42 per cent above 1990 emissions by 2010.[2] Escalating petrol prices are further exacerbating the demand for technological change in the industry to more fuel-efficient or fuel-alternative vehicles.

The reluctance for change from the automotive industry is partly a result of adopting traditional economic theory; assuming that a major improvement in fuel efficiency or emissions reduction must be traded off against cost, performance or safety – thereby making it an unattractive option for both the manufacturer and the consumer.

A number of technological innovations have emerged over the last decade that fly in the face of the environment vs. economy trade-off and have challenged the conventional approach to automobile design and manufacture. One of the most impressive innovations is the Hypercar Revolution concept, conceived by the team from the Rocky Mountain Institute and its partners, led by Amory B. Lovins. The Revolution vehicle uses existing technologies with Whole System Design (WSD) to show that:

- Very large improvements in fuel economy and carbon emissions may be easier and cheaper than small ones, and may not conflict with existing performance objectives;
- These improvements may bring about competitive advantage to manufacturers by reducing costs and requirements associated with capital, assembly, space, parts and product takeback;
- A business model based on value to the customer is an advantage to manufacturers; and
- Vehicle production and use will become less susceptible to fuel price, government policy and other variable pressures.

Question: Is it technically possible and economically advantageous to both the manufacturer and the consumer to design high-performance vehicles that produce less toxic emissions, use clean or existing fuels more efficiently, and deliver more value?

Answer: Yes. A case in point is the Toyota Prius. Since its launch in Australia in October 2003, the hybrid has proved a hot item among private and business owners. With fuel economy at 4.4l/100km and a suite of new technologies, the Prius is a prime example of the change that is needed – and is now happening – in the automotive industry.

Worked example overview

Cars are a subgroup of powered road vehicles. Other subgroups include trucks, buses and motorcycles. These vehicles may incorporate any of a number of power plants, including combustion power plants of petrol/gasoline, LPG, diesel or biofuels; hybrid-electric power plants; battery-electric power plants; and fuel cell power plants. Further information about the Whole System Design of cars and trucks with various power plants is available in The Natural Edge Project's freely available online textbook *Energy Transformed: Sustainable Energy Solutions for Climate Change Mitigation*, 'Lecture 8.2: Integrated approaches to energy efficiency and alternative transport fuels – passenger vehicles' and 'Lecture 8.3: Integrated approaches to energy efficiency and alternative transport fuels – trucking'.[3] The following worked example focuses on cars with fuel cell power plants.

Recall the ten elements of applying a WSD approach discussed in Chapters 4 and 5:

1. Ask the right questions;
2. Benchmark against the optimal system;
3. Design and optimize the whole system;
4. Account for all measurable impacts;
5. Design and optimize subsystems in the right sequence;
6. Design and optimize subsystems to achieve compounding resource savings;
7. Review the system for potential improvements;
8. Model the system;
9. Track technology innovation; and
10. Design to create future options.

The following worked example will demonstrate how the ten elements can be applied to passenger vehicles using two contrasting examples: a conventional passenger vehicle versus the Hypercar Revolution concept, developed by the Rocky Mountain Institute. The main focus is on elements 3, 4 and 5. The application of the other elements will be indicated with a shaded box.

Primarily, the following vehicle subsystems will be considered:

- Structure;
- Propulsion;
- Chassis;
- Electrical;
- Trim; and
- Fluids.

Design challenge

Design a passenger vehicle. Make it better than the previous model.

Design process

Both the conventional vehicle and WSD Hypercar Revolution examples will consider the following steps with respect to the development process, each of which is correlated with an element of WSD, first introduced by the Rocky Mountain Institute:

1. *Design*: Determining the general component composition for each subsystem (Element 3);
2. *Optimization*: Making the vehicle the best it can be based on the general component composition (Element 5); and
3. *Cost analysis*: Comparison of the economic and environmental costs and benefits (Element 4).

Vehicle design

Conventional vehicle design

The lack of variation in vehicle subsystems over the last 70 years is an indication that the conceptual design phase of producing passenger vehicles has involved little more than simply copying a previous design.

The subsystems of almost every passenger vehicle produced in the last 70 years have the following characteristics and components:

- *Structure*: Made from steel due to low cost and high stiffness and durability; expensive to produce due to high tooling costs; components designed for manufacturability.

- *Propulsion*: Components include internal combustion engine (ICE), starter motor, alternator, radiator, transmission, driveshaft, differential, axles and fuel tank; mostly mechanical power transmission; no energy recovery; mostly made from steel due to high strength and ductility; expensive to produce due to high tooling costs; fuel source is usually petroleum, sometimes diesel or LPG.
- *Chassis*: Components include suspension, braking, wheels and steering systems; mechanical function; no energy recovery; sized to support structure and propulsion systems.
- *Electrical*: Components include heating, venting and air conditioning (HVAC), lighting and audio systems; most components are linked with designated point-to-point wiring.
- *Trim*: Fabrics cover the entire structure in the passenger cabin; style based on ergonomics, fashion and safety.
- *Fluids*: Components include a water system for the HVAC and cooling the ICE; fuel in fuel lines; transmission fluid; brake fluid; oil and grease for lubrication of mechanical moving parts.

The subsystems are integrated to produce a passenger vehicle only *after* the decisions are made as to their general component composition, as in Figure 7.1.

Whole System Design

Element 3: Design and optimize the whole system

In designing the vehicle, Element 3: 'Design and optimize the whole system', can be interpreted as 'Design the system as a whole' by considering the effect of each subsystem on every other subsystem in an integrated manner, as in Figure 7.2.

To explore the potential technological options, the Hypercar team started the design process with a clean sheet.[4] They removed almost all constraints (political and material) associated with conventional passenger vehicle design and focused on designing the Revolution as a whole system. This process – known as cleansheet design – provides scope for emphasis on two key features of the conceptual design phase:

1. Designing the future into the system using a process known as *backcasting*, which makes the benefits of the system *more attractive* with technological and political progress; and
2. Designing components to provide *multiple services*, hence introducing system synergies that help overcome conventional compromises and lead to multiple, compounding benefits.

Backcasting

Element 10: Design to create future options

Several features of the Revolution, including the use of advanced composite materials, hydrogen fuel cells, the Fiberforge™ manufacturing process, and extensive

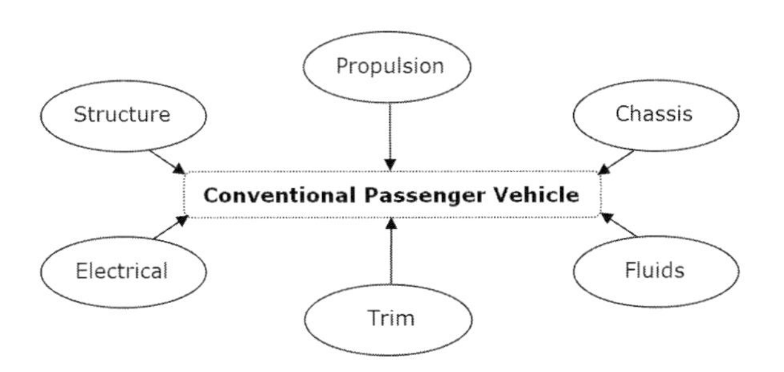

Figure 7.1 *The component optimization strategy of conventional vehicle design*

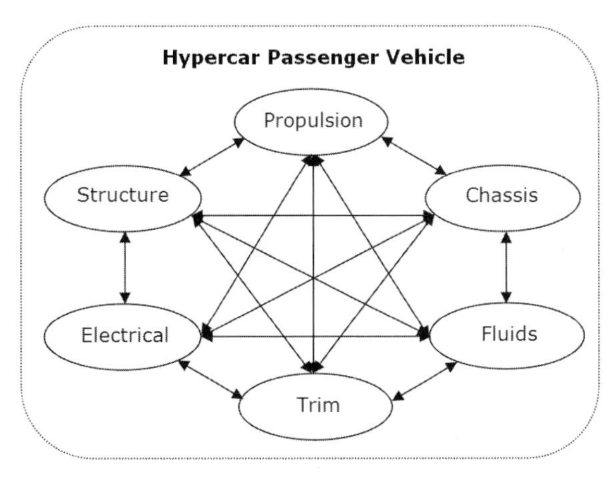

Figure 7.2 *The system design strategy of Whole System vehicle design*

electronic and software control, are a result of backcasting, as in Figure 7.3. These features were selected because they are currently viable steps *on the path* to a sustainable vehicle.

The Revolution was designed to accommodate technological and political progress. As a result, the backcast upgrades to the Revolution, unlike the forecast upgrades to a conventional vehicle, are cheap and easy, and are subject to *expanding returns* instead of *diminishing returns*.

Designing for multiple services

Generally, conventional cars are either 'small, light, clean and efficient' or 'large, powerful, comfortable and safe' – anything else involves compromise. However, these compromises do not exist until they are designed into the vehicle. The Hypercar team used clean-sheet design and WSD to create a vehicle that is light, clean, efficient, spacious, well-performing, comfortable and safe. Many conventional compromises were avoided by:

- Designing components for multiple purposes, which results in compounding benefits throughout the Revolution; and
- Employing several advanced technologies that are currently not cost-effective on their own but become cost-effective when combined.

Every vehicle subsystem comes into direct physical contact with the structure subsystem. Thus there is an opportunity to eliminate some complexity in the other dynamic subsystems by simplifying their integration with the structure or by transferring some of their functions and services to the structure. For example, the Revolution's structure includes suspension system mounts, cooling lines, electrical conduits and trim features,[5] which lead to compounding benefits with the corresponding subsystems:

- The cooling lines are part of a single-circuit cooling system that moderates the temperature of several components.[6] In contrast, conventional vehicles usually have dedicated cooling circuits for each component, which are heavier, more complex and more expensive.
- The electrical conduits are part of a network-based electronic control system that replaces conventional, dedicated, point-to-point wiring and control. This all-in-one system reduces cost, complexity, failure modes and diagnostic problems.[7]
- The structure's trim features are exposed to some parts of the passenger cabin, where they double as cosmetic trim.[8] The trim features simplify and reduce the cost of the remaining trim.

The front end of the structure is mostly made from aluminium and provides two services to reduce complexity, mass and cost. The first service is crash-resistance through energy absorption. The second service is as housing for the front end components of the propulsion system, but without the complexity of conventional, add-on, mounting structures.[9]

The rest of the Revolution's structure is primarily made from advanced composite materials. The advantages of the composite materials, from which

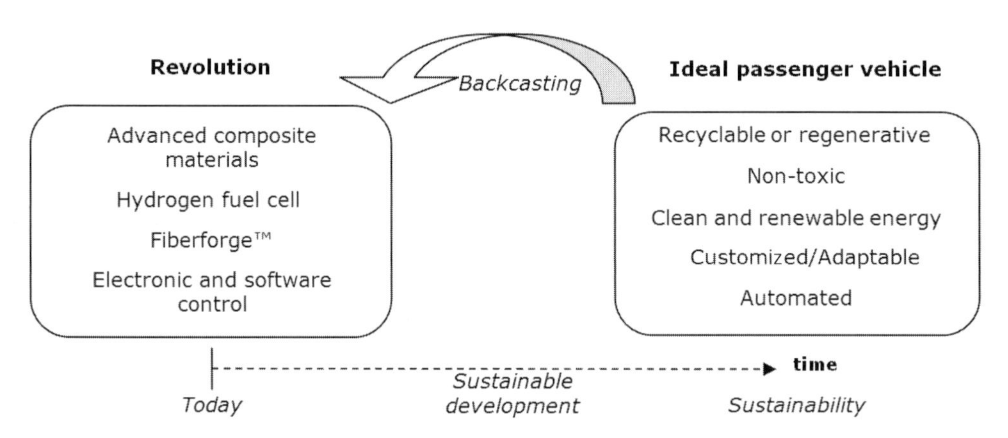

Figure 7.3 *Selecting vehicle components after backcasting from an ideal sustainable vehicle*

several compounding benefits are initiated, are that they are lightweight, stiff, have a low thermal mass and provide acoustic insulation:[10]

- The low weight and high stiffness of the composite structure make the hydrogen fuel cell-based propulsion system, which is quiet, energy-efficient and virtually emissions-free, cost-effective. Fuel cells typically require a radiator about twice the size of that of equivalent conventional internal combustion engines when retrofitted into existing cars. However, the low mass of the Revolution's structure can lead to a reduction in required propulsion power by a factor of up to five, reducing the size of the fuel cell, which means that overall the radiator is smaller than that of a conventional vehicle.[11]
- The fuel cell power is supplemented with battery power, which is used to meet peak loads, such as during acceleration, towing and driving up an incline. The batteries are partly charged by the braking system, which captures braking energy. Since passenger vehicles usually brake as much as they accelerate, the batteries rarely need charging.
- The low thermal mass of the structure also provides a thermal insulation service, which, with the integrated cooling lines, helps simplify and quieten the climate control system. The Revolution's windows also assist with climate control. The windows employ spectrally selective glazing, which reduces infrared (heat) gain in the passenger cabin.[12]
- The quiet fuel cell/electric system and climate control system, coupled with acoustic insulation provided by the composite structure, make for a more comfortable ride.

Only about 7.5 per cent of the mass of a conventional vehicle is from composite or plastic materials, most of which are for non-structural components.[13] The conventional opinion is that producing composite components involves high labour and hence high cost. However, the Hypercar team used clean-sheet design to devise a method of cost-effectively producing the Revolution's extensive catalogue of composite parts. In fact, the setup costs are a fraction of tooling costs for conventional steel vehicles.[14]

The advantages of the production method are a result of multiple benefits and synergies between material selection, subsystem design and a production process called Fiberforge™:

- The composite materials produced for the Revolution contain about 55–65 per cent fibre material, compared with 20–30 per cent for conventional composite materials.[15] The fibre material is responsible for most of the strength and stiffness of the composite, which means that the Revolution's composite can get the same performance with less mass and cost than the conventional composite. The Revolution's composite uses long discontinuous fibre (LDF) carbon, which, compared to using continuous fibres, improves stretchability, processability and formability with very little loss in strength and stiffness.[16]
- The Revolution's structure subsystem consists of only 14 major parts – about 65 per cent fewer than that of a conventional structure – and only 62 total parts – about 77 per cent fewer than that of a conventional structure,[17] which simplifies assembly and reduces effort by about 80 to 90 per cent.[18] Although some parts have complex surface geometry, all are shallow and few require sharp bends or deep draws, which increases repeatability and eliminates the need for additional clean-up steps.[19]
- Fiberforge™ is a flexible, software-intensive production process that incorporates relatively few steps and tailored blanks. Tailored blanks are composite sheets that are roughly the desired shape and have the correct fibre alignment, angle and thickness for the final component. They eliminate the need for additional assembly and processing steps. The use of tailored blanks also results in simpler assembly and only 15 per cent material scrap,[20] compared with about 30–40 per cent for conventional stamped steel processing,[21] saving on both material and clean-up costs.

The Revolution's bare exterior panels also provide multiple services. Aside from providing protection and aerodynamic streamlining, they also provide aesthetic appeal. Panel production for the Revolution includes in-mould colouring, which eliminates the need for extra preparation and painting steps.[22]

Vehicle optimization

Conventional vehicle optimization

Conventional vehicle design uses incremental product refinement for each subsystem in isolation to improve component quality, introduce features and enhance manufacturing efficiency with current production systems, as in Figure 7.4.

The following incremental improvements are typically made to subsystems:

- *Structure*: Incremental improvements made to structure without having to significantly modify tooling (changing the tooling is often too expensive to justify change; hence the previous model structure is used).
- *Propulsion*: Minor adjustments such as valve timing, improving small components and tolerances, adding components such as cams, improving power output from the engine (by 5–10 per cent), and increasing fuel consumption through power improvements or a bigger engine.
- *Chassis*: Adjust the suspension to accommodate larger engine mass; larger brakes to accommodate faster speed and acceleration; steering and wheel systems are unaffected; use previous model steering and wheel systems.
- *Electrical*: Adjust electrical and electronic systems corresponding to modifications made to other subsystems such as propulsion and chassis.
- *Trim*: Select a new colour scheme based on popular fashion trends.
- *Fluids*: Larger water hoses to cool a more powerful engine; larger fuel lines to feed a more powerful engine; larger fuel tank for higher fuel consumption; larger brake fluid tank and lines for larger brakes.

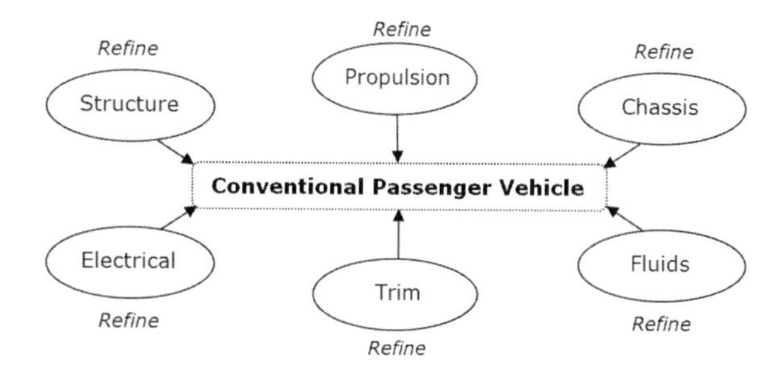

Figure 7.4 *The component optimization strategy of conventional vehicle design*

As a result of the above incremental improvements:

1. The vehicle is slightly faster, more powerful and more attractive to forecasted market trends;
2. The vehicle mass is variant – slightly lighter, heavier or negligibly different; and
3. Vehicle fuel consumption is variant – slightly less, more or negligibly different.

WSD optimization

> *Element 5: Design and optimize subsystems in the right sequence*

A WSD approach to optimization involves taking advantage of *synergies* between components and measuring modifications with respect to the whole system.

A primary performance factor of vehicle design is *fuel consumption per unit of distance*, B_e, and can be calculated using Equation 7.1 and Table 7.1.

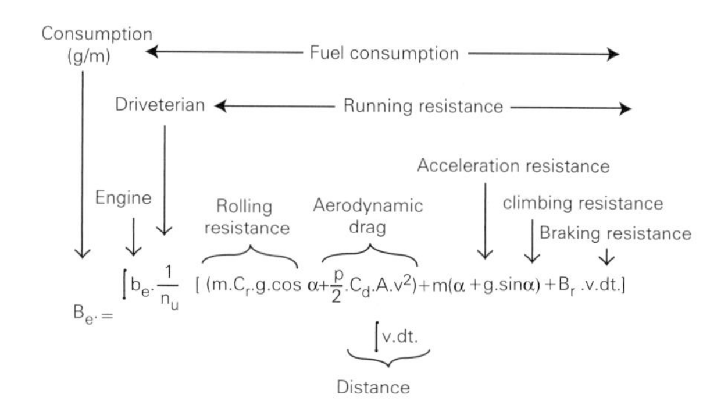

$$B_e = \frac{\int b_e \cdot \frac{1}{n_u} \left[(m.C_r.g.\cos\alpha + \frac{p}{2}.C_d.A.v^2) + m(\alpha + g.\sin\alpha) + B_r .v.dt. \right]}{\int v.dt.}$$

Source: Robert Bosch GmbH (2004)[23]

Equation 7.1 *Fuel consumption per unit of distance for road vehicles*

> *Element 7: Review the system for potential improvements*

The above equation suggests that fuel consumption is most influenced by engine efficiency (b_e) and drivetrain efficiency (η_{ii}). Hence the emphasis has been on technologies such as fuel cells and hybrid drives as key to improving fuel-efficiency. While this relationship may be true in the case of incremental product refinement, it

Table 7.1 *Symbol nomenclature*

Quantity	Unit	Quantity	Unit
B_e Consumption per unit of distance	g/m	C_d Coefficient of aerodynamic drag	–
$\eta_{ü}$ Transmission efficiency of drivetrain	–	A Frontal area	m^2
m Vehicle mass	kg	v Vehicle speed	m/s
C_r Coefficient of rolling resistance	–	a Acceleration	m/s^2
g Gravitational acceleration	m/s^2	B_r Braking resistance	N
α Angle of ascent	°	t Time	s
ρ Air density	kg/m^3	b_e Specific fuel consumption	g/kWh

Source: Robert Bosch GmbH (2004)[24]

does not take into account the benefits of 'mass decompounding' or other subsystem improvements. As the equation above suggests, there are a number of additional factors to consider in improving fuel economy:

- rolling resistance;
- aerodynamic drag;
- acceleration resistance;
- climbing resistance; and
- breaking resistance.

Mass decompounding – Reducing mass first

Element 6: Design and optimize subsystems to achieve compounding resource savings

Mass (m) is directly proportional to rolling resistance, acceleration resistance and climbing resistance, which are all used to determine required peak power of vehicles. Reducing platform mass means reducing fuel cell size and cost. 'Mass decompounding' is the term used to describe the snowballing of weight savings, or a 'beneficial mass spiral'. A lighter vehicle body requires a lighter chassis and smaller powertrain, which further reduces mass. A number of mass reduction iterations can lead to much lighter components, or even the elimination of some components. For example, installing a hybrid-electric system would remove the need for a transmission, clutch, flywheel, starter motor and alternator, among other components. Although the new system would require the installation of a battery, power electronics and driver motor, the result may be a net reduction in mass and cost. Irrespective of the net change, any powertrain will be smaller and more cost-effective in a lighter vehicle platform than a heavier one.

Element 8: Model the system

Figure 7.5 indicates the synergies between components in the Revolution with respect to mass. We can use the synergies to our advantage by applying Element 5: 'Design and optimize subsystems at the right time and in the right sequence'. From Figure 7.5 we can see that the mass of the structure should be minimized first. Doing so reduces the need for a large propulsion system and large amounts of trim. A lighter structure and smaller propulsion system means that the chassis can also be made lighter. Similarly, a smaller propulsion system and lighter chassis means that the electrical systems can be made lighter. Finally, a smaller propulsion system requires a smaller volume of fluids. This strategy for optimizing the Revolution will maximize the mass reductions for the least amount of effort.

Structure

Element 9: Track technology innovation

In order to make a component lighter, it makes sense to investigate lighter materials. Designers are conventionally encouraged to optimize *component* functionality for *minimum* cost, and so steel is usually

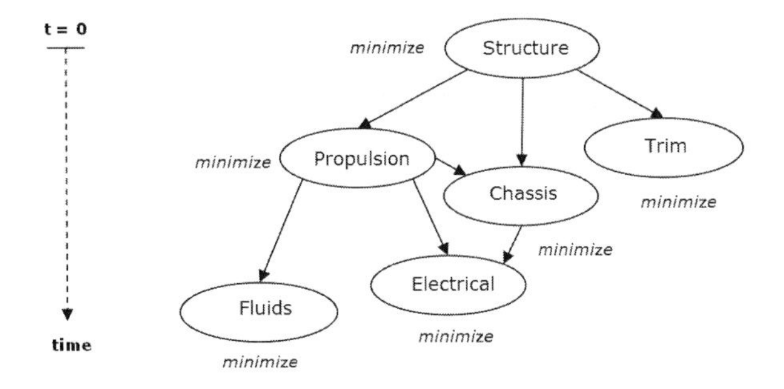

Figure 7.5 *The flow of compounding mass reduction in the system design strategy of Whole System vehicle design*

selected. However, a comparably priced automobile structure can be created using advanced composite materials (primarily carbon/epoxy) and lightweight metals such as aluminium and magnesium alloys:[25]

- The advanced composite has a much higher stiffness-to-density ratio than steel, so less mass is required for the same stiffness;
- The advanced composite can be processed and assembled more easily, with low scrap (15 per cent compared to 30–40 per cent for steel)[26] and at a fraction of the cost of steel;[27] and
- The advanced composite and alloys can be recycled with negligible loss to integrity, and therefore have a high salvage value.[28]

Using mostly advanced composite materials and some aluminium, the Revolution's complete structure has a mass of only 187kg (57 per cent less than that of a conventional steel structure), and it also includes features such as integrated suspension mounting, cooling lines, electrical conduits and trim features, which result in compounding mass savings with the corresponding subsystems.[29] The composite materials have a low thermal mass, which makes them good insulators and reduces the need for additional insulating materials.

Propulsion

By optimizing the structure first, the mass reductions can be carried over to the propulsion system. The reduction would normally lead to a roughly 57 per cent reduction in power required to move the structure. Assuming that the conventional propulsion system generates 180kW of output power, which is typical of a modern 6-cylinder sedan, we would expect the Revolution to require a 75–80kW propulsion system. The low power requirement makes some currently expensive technologies viable. A particularly favourable option is a hydrogen fuel cell. In fact, the fuel cell-powered Revolution requires only 35kW for cruising and up to 70kW during peak loads,[30] such as during acceleration, towing and driving up an incline.

The lower than expected power output is attributable to four factors:

1. Not only is the structure lighter, but the propulsion system is lighter too. The fuel cell system has a mass of only 288kg, 38 per cent lower than a conventional propulsion system.[31] Therefore, lower power output is required to move the lower mass.
2. The hydrogen fuel cell system is more fuel efficient (29 per cent efficient) than the conventional internal combustion engine (15–20 per cent efficient).[32] Therefore more fuel is used to move the vehicle and less is being lost through friction and heat.
3. The design of the structure and propulsion systems result in 55 per cent lower aerodynamic drag.[33]
4. The design of the propulsion and chassis systems result in 65 per cent lower rolling resistance.[34]

Chassis

The lighter structure and propulsion systems reduce the suspension, braking and steering loads, which leads to reduced chassis mass. The chassis mass is reduced further through features integrated into the structure and electrical propulsion systems, and through integrated software control, which eliminates the need for several mechanical components:[35]

- The suspension system in the Revolution is lighter than that of a conventional vehicle since it does not need to support as much mass;
- The braking system in the Revolution is also lighter than that of a conventional car since it does not need to decelerate as much mass;
- The PAX run-flat tyre system was developed for the Revolution; this has a 15 per cent lower rolling resistance than conventional tyre systems.[36]

The steering mechanism in the Revolution is electrically actuated and lighter than a conventional steering column.

Electrical

The Revolution's electrical and electronic control system is a low-cost, network-based, shared-data system.[37] The network structure eliminates the need for designated, point-to-point wiring as in a conventional vehicle. As a result, the mass of the system is reduced. The electrical accessories, including heating, venting and air-conditioning (HVAC), lighting, and audio systems, are four times more energy-efficient than the conventional systems,[38] which means that they emit less heat and thus reduce the required load on the HVAC system.

Trim

Since the trim features of the Revolution's structure double as interior surfaces for many parts of the passenger cabin, the need for conventional cosmetic trim is reduced.[39]

Fluids

The intensive electrical integration in the Revolution, particularly in the propulsion system, eliminates the need for many of the fluids that the conventional mechanical system requires for cooling and lubrication. The Revolution does not require transmission fluid, brake fluid or oil for engine lubrication, and requires less water for cooling and less grease for lubrication.[40] In addition to the reduced fluids requirement, the Revolution also needs to carry far less fuel than a conventional vehicle. The Revolution has a range of 530km on 3.4kg of hydrogen,[41] whereas a conventional car has a range of about 450km on 53kg, or 72l, of petrol. This difference equates to a 95 per cent improvement in mass efficiency in favour of the Revolution.

Result

The total mass of the vehicle is only 857kg (52 per cent less mass than a conventional vehicle of the same size) and it is roughly the same size and price as a conventional vehicle (Figure 7.6 shows the subsystem breakdown). Furthermore, the Revolution emits zero emissions with equal or better performance than a conventional vehicle.

Cost analysis

Conventional cost analysis

The price of a conventional business-class sedan is usually US$50,000–$70,000. This price does not include the cost of regular servicing and replacement parts, which can amount to US$500–$2000 per year depending on the parts and level of service. In fact, some vehicles are sold at an initial loss, with the expectation

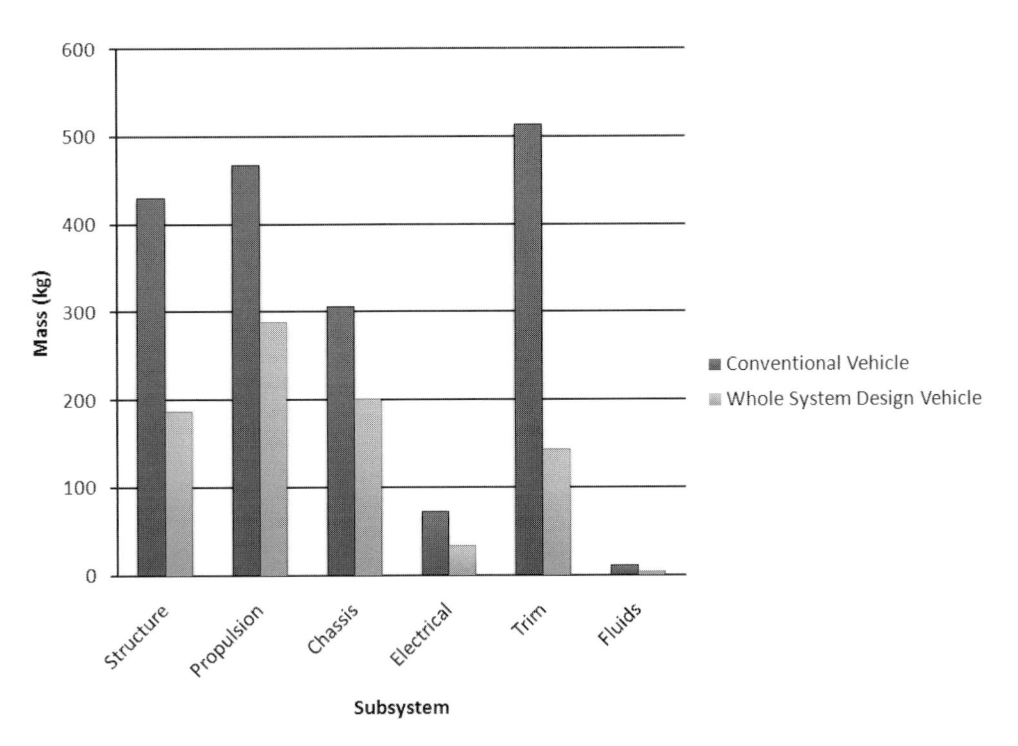

Figure 7.6 *Mass comparison between the conventional passenger vehicle and the WSD vehicle, by subsystem*

Table 7.2 *Average life of some serviceable components in a conventional passenger vehicle*

Component	Average Life	Component	Average Life
Transmission fluid	30,000km	Brake components	5 years
Oil	6 months		
Oil filter	6 months	Brake fluid	15,000km
Spark plugs	10,000km	Tyres	50,000km
Coolant	2 years	Water	6 months

that a profit will be made through aftermarket servicing and the sale of replacement parts, as in Table 7.2, over the roughly 15-year, 300,000km life of the vehicle. However, not included in Table 7.2 are the unserviceable components that usually fail before the end of the vehicle's life, such as the clutch, ball joints, muffler and gaskets. These components are generally more expensive than the serviceable components in Table 7.2.

A conventional business-class passenger vehicle has a fuel economy of about 8km/litre. At 2005 petrol prices of about US$1.30/litre at the pump, fuel costs for such a vehicle are about 16c/km. It is likely that this figure will climb with rising petrol prices.

WSD cost analysis

Element 4: Account for all measurable impacts

The total ongoing production cost of the Revolution is similar to that of a conventional vehicle. However, the cost of some technologies, such as fuel cell technology, is falling. Although upfront design effort for the Revolution is greater, the overall capital investment is several times smaller than that of a conventional vehicle, which is dominated by tooling costs.[42] Tooling for conventional vehicles usually involves designated dyes and moulds for each of many components, whereas the Revolution's Fiberforge™ system requires relatively few tools to make all of its advanced composite components.[43] Therefore the Revolution can be sold at a similar price as the conventional vehicle, but at a profit. Furthermore, economic justification need not depend on aftermarket sales.

Aside from owning a better vehicle, the customer also enjoys lower aftermarket costs:

- Fewer components require replacement during the Revolution's 300,000km or more service life.[44] Extensive electronic control has eliminated the need for many mechanical components that usually wear out. Even the components that are similar to that of a conventional vehicle, such as brake calipers and rotors,[45] should last the life of the Revolution. Fewer mechanical parts means less and cleaner fluids, so fluid replacement and service cost is reduced.
- The Revolution has a fuel economy of about 156km/kg.[46] At 2003 hydrogen prices of US$3.51/kg,[47] fuel costs are about 2c/km. It is likely that this figure will only fall, with the development of hydrogen-extraction technologies. By applying Element 4, 'Account for all measurable impacts', the customer may also enjoy lower costs for other services in the future:
- Governments may offer financial incentives for purchasing a Revolution, such as a rebate on purchase cost or a built-in discount on other taxes such as lower vehicle registration costs.
- The Revolution may be cheaper than a conventional vehicle to insure, as its panels are resistant to minor dents and scratches,[48] it is safer so the likelihood of passenger harm is lower, and its low mass will inflict less damage on another car during a collision.
- In a future hydrogen economy, a parked Revolution can be used to generate electricity for other purposes. Plug-in systems, similar but with opposite flow to those used to charge the batteries of electric vehicles, can transfer electricity from the Revolutions to local buildings to help meet energy demands. Electricity can be sold to the grid, generating income for the owner.[49]

The Revolution also boasts some impressive environmental benefits that are generally not counted by original equipment manufacturers (OEMs) or customers:

- The Revolution generates zero emissions.[50] The only by-product from the fuel cell is heat and pure water, which can be collected and put to use

Table 7.3 *Some environmental impacts of a conventional vehicle and the Hypercar Revolution*

Waste	Mass (kg)	Waste fraction	Material	Direct waste (kg)	Hidden abiotic waste (kg)	Hidden water waste (t)
Conventional vehicle						
Structure scrap	430	35%	Steel[54]	151	1279	11.3
Non-recycled	1800	25%	Glass, plastic, rubber[55]	450	1800	127.8
Total				*601*	*3079*	*139.1*
Hypercar Revolution						
Structure scrap	187	15%	Carbon fibre[56]	28	1629	50.5
Non-recycled	857	0%	n/a	0	0	0
Total				*28*	*1629*	*50.3*

elsewhere. Zero emissions also means cleaner air and buildings, reduced health risks for people, fewer regulatory restrictions, and simplified testing for OEMs.

- The quiet operation of the fuel cell/electric propulsion system reduces noise pollution.
- Revolution also has a high salvage value since it is almost fully recyclable.[51] Even the valuable carbon fibre material, which conventionally would be sent to landfill, can be recovered profitably with negligible degradation.

The true value of recovering materials emerges when we investigate their *hidden* costs. For example, the production of 1kg of aluminium requires 19kg of other abiotic materials and 539kg of water.[52] Although aluminium and the advanced composite (carbon fibre and epoxy resin) of the Revolution have a greater environmental impact per unit mass than the steel of the conventional vehicle, they are recovered and reused, so their environmental impact is compensated for by their providing more services throughout their life. On the other hand, the roughly 25 per cent conventional vehicle mass that is landfilled mainly consists of the non-metallic components – glass, plastics (such as nylon, polypropylene and polyvinyl chloride) and rubber.[53] The environmental impact from the loss of these single-use materials puts the Revolution at an even greater environmental advantage. Table 7.3 compares the environmental impact of the scrap material generated when building the vehicles' structures, and the impact of the non-recycled materials that are landfilled at the end of the vehicles' lives. Comparing just the structure scrap, the Revolution performs worse than the conventional vehicle. However, through careful selection of the Revolution's materials and processing techniques, some of which are more expensive than those of the conventional vehicle, the Revolution is almost fully recyclable. Many of the Revolution's more expensive options, like the use of carbon fibre for the structure, resulted in increased recyclability and hence overall smaller environmental impact when compared to the conventional vehicle.

Notes

1 See Australian Greenhouse Office, 'Sustainable transport', at www.greenhouse.gov.au/transport/, accessed 14 April 2006.
2 See Australian Greenhouse Office, 'Sustainable transport', at www.greenhouse.gov.au/transport/, accessed 14 April 2006.
3 Smith, M., Hargroves, K., Stasinopoulos, P., Stephens, R., Desha, C. and Hargroves, S. (2007) *Energy Transformed: Sustainable Energy Solutions for Climate Change Mitigation*, The Natural Edge Project, Australia, 'Lecture 8.2: Integrated approaches to energy efficiency and alternative transport fuels – passenger vehicles', and 'Lecture 8.3: Integrated approaches to energy efficiency and alternative transport fuels – trucking', www.naturaledgeproject.net/Sustainable_Energy_Solutions_Portfolio.aspx, accessed 10 April 2008.
4 Lovins, A. B. and Cramer, D. R. (2004) 'Hypercars, hydrogen, and the automotive transition', *International Journal of Vehicle Design*, vol 35, nos 1–2, pp50285.
5 Cramer, D. R. and Taggart, D. F. (2002) 'Design and manufacture of an affordable advanced-composite

automotive body structure', in *Proceedings of the 19th International Battery, Hybrid and Fuel Cell Electric Vehicle Symposium and Exhibition*, www.rmi.org/images/other/Trans/T02-10_DsnManuAdvComp.pdf, accessed 25 April 2007.

6 Lovins, A. B. and Cramer, D. R. (2004) 'Hypercars, hydrogen, and the automotive transition', *International Journal of Vehicle Design*, vol 35, nos 1–2, pp50–85, www.rmi.org/images/other/Trans/T04-01_Hypercar H2AutoTrans.pdf, accessed 25 August 2008.

7 Lovins, A. B. and Cramer, D. R. (2004) 'Hypercars, hydrogen, and the automotive transition', *International Journal of Vehicle Design*, vol 35, nos 1–2, pp50–85, www.rmi.org/images/other/Trans/T04-01_Hypercar H2AutoTrans.pdf, accessed 25 August 2008.

8 Cramer, D. R. and Taggart, D. F. (2002) 'Design and manufacture of an affordable advanced-composite automotive body structure', in *Proceedings of the 19th International Battery, Hybrid and Fuel Cell Electric Vehicle Symposium and Exhibition*, www.rmi.org/images/other/Trans/T02-10_DsnManuAdvComp.pdf, accessed 25 April 2007.

9 Lovins, A. B. and Cramer, D. R. (2004) 'Hypercars, hydrogen, and the automotive transition', *International Journal of Vehicle Design*, vol 35, nos 1–2, pp50–85, www.rmi.org/images/other/Trans/T04-01 Hypercar H2AutoTrans.pdf, accessed 25 August 2008.

10 Cramer, D. R. and Taggart, D. F. (2002) 'Design and manufacture of an affordable advanced-composite automotive body structure', in *Proceedings of the 19th International Battery, Hybrid and Fuel Cell Electric Vehicle Symposium and Exhibition*, www.rmi.org/images/other/Trans/T02-10_DsnManuAdvComp.pdf, accessed 25 April 2007.

11 Lovins, A. B. and Cramer, D. R. (2004) 'Hypercars, hydrogen, and the automotive transition', *International Journal of Vehicle Design*, vol 35, nos 1–2, pp50–85, www.rmi.org/images/other/Trans/T04-01_Hypercar H2AutoTrans.pdf, accessed 25 August 2008.

12 Lovins, A. B. and Cramer, D. R. (2004) 'Hypercars, hydrogen, and the automotive transition', *International Journal of Vehicle Design*, vol 35, nos 1–2, pp50–85, www.rmi.org/images/other/Trans/T04-01_Hypercar H2AutoTrans.pdf, accessed 25 August 2008.

13 Ward Communications (1999) cited in Lovins, A. B. and Cramer, D. R. (2004) 'Hypercars, hydrogen, and the automotive transition', *International Journal of Vehicle Design*, vol 35, nos 1–2, pp50285, www.rmi.org/images/other/Trans/T04-01_HypercarH2AutoTrans.pdf, accessed 25 August 2008.

14 Williams, B. D. et al (1997) *Speeding the Transition: designing a fuel-cell Hypercar*, Rocky Mountain Institute, Colorado, www.rmi.org/images/other/Trans/T97-09_SpeedingTrans.pdf, accessed 14 January 2005.

15 Cramer, D. R. and Taggart, D. F. (2002) 'Design and manufacture of an affordable advanced-composite automotive body structure', in *Proceedings of the 19th International Battery, Hybrid and Fuel Cell Electric Vehicle Symposium and Exhibition*, www.rmi.org/images/other/Trans/T02-10_DsnManuAdvComp.pdf, accessed 14 January 2007.

16 Cramer, D. R. and Taggart, D. F. (2002) 'Design and manufacture of an affordable advanced-composite automotive body structure', in *Proceedings of the 19th International Battery, Hybrid and Fuel Cell Electric Vehicle Symposium and Exhibition*, www.rmi.org/images/other/Trans/T02-10_DsnManuAdvComp.pdf, accessed 14 January 2007.

17 Lovins, A. B. and Cramer, D. R. (2004) 'Hypercars, hydrogen, and the automotive transition', *International Journal of Vehicle Design*, vol 35, nos 1–2, pp50–85, www.rmi.org/images/other/Trans/T04-01_HypercarH2AutoTrans.pdf, accessed 25 August 2008.

18 Lovins, Brylawski, Cramer and Moore (1996) cited in Fox, J. W. and Cramer, D. R. (1997) *Hypercars: A Market-Oriented Approach to Meeting Lifecycle Environmental Goals*, Rocky Mountain Institute, Colorado, www.rmi.org/images/other/Trans/T97-05_MarketApproach.pdf, accessed 8 June 2005.

19 Lovins, A. B. and Cramer, D. R. (2004) 'Hypercars, hydrogen, and the automotive transition', *International Journal of Vehicle Design*, vol 35, nos 1–2, pp50–85, www.rmi.org/images/other/Trans/T04-01_Hypercar H2AutoTrans.pdf, accessed 25 August 2008.

20 Cramer, D. R. and Taggart, D. F. (2002) 'Design and manufacture of an affordable advanced-composite automotive body structure', in *Proceedings of the 19th International Battery, Hybrid and Fuel Cell Electric Vehicle Symposium and Exhibition*, www.rmi.org/images/other/Trans/T02-10_DsnManuAdvComp.pdf, accessed 14 January 2007.

21 Fox, J. W. and Cramer, D. R. (1997) *Hypercars: A Market-Oriented Approach to Meeting Lifecycle Environmental Goals*, Rocky Mountain Institute, Colorado, www.rmi.org/images/other/Trans/T97-05_MarketApproach.pdf, accessed 8 June 2005.

22 Cramer, D. R. and Taggart, D. F. (2002) 'Design and manufacture of an affordable advanced-composite automotive body structure', in *Proceedings of the 19th International Battery, Hybrid and Fuel Cell Electric Vehicle Symposium and Exhibition*, www.rmi.org/images/other/Trans/T02-10_DsnManuAdvComp.pdf, accessed 14 January 2007.

23 Robert Bosch GmbH (2004) *Automotive Handbook*, Robert Bosch GmbH, Warrendale, PA, US, p417.

24 Robert Bosch GmbH (2004) *Automotive Handbook*, Robert Bosch GmbH, Warrendale, PA, US, p417.

25 Research is currently being undertaken by CSIRO on lightweight metals for cars (www.csiro.au/csiro/content/standard/ps13x,,.html).
26 Cramer, D. R. and Taggart, D. F. (2002) 'Design and manufacture of an affordable advanced-composite automotive body structure', in *Proceedings of the 19th International Battery, Hybrid and Fuel Cell Electric Vehicle Symposium and Exhibition*, www.rmi.org/images/other/Trans/T02-10_DsnManuAdvComp.pdf, accessed 14 January 2007.
27 Williams, B. D. et al (1997) *Speeding the Transition: Designing a Fuel-Cell Hypercar*, Rocky Mountain Institute, Colorado, www.rmi.org/images/other/Trans/T97-09_SpeedingTrans.pdf, accessed 14 January 2005.
28 Fox, J. W. and Cramer, D. R. (1997) *Hypercars: A Market-Oriented Approach to Meeting Lifecycle Environmental Goals*, Rocky Mountain Institute, Colorado, www.rmi.org/images/other/Trans/T97-05_MarketApproach.pdf, accessed 8 June 2005.
29 Cramer, D. R. and Taggart, D. F. (2002) 'Design and manufacture of an affordable advanced-composite automotive body structure', in *Proceedings of the 19th International Battery, Hybrid and Fuel Cell Electric Vehicle Symposium and Exhibition*, www.rmi.org/images/other/Trans/T02-10_DsnManuAdvComp.pdf, accessed 14 January 2007.
30 Lovins, A. B. and Cramer, D. R. (2004) 'Hypercars, hydrogen, and the automotive transition', *International Journal of Vehicle Design*, vol 35, nos 1–2, pp50–85, www.rmi.org/images/other/Trans/T04-01_HypercarH2AutoTrans.pdf, accessed 25 August 2008.
31 Cramer, D. R. and Taggart, D. F. (2002) 'Design and manufacture of an affordable advanced-composite automotive body structure', in *Proceedings of the 19th International Battery, Hybrid and Fuel Cell Electric Vehicle Symposium and Exhibition*, www.rmi.org/images/other/Trans/T02-10_DsnManuAdvComp.pdf, accessed 14 January 2007.
32 Williams, B. D. et al (1997) *Speeding the Transition: Designing a Fuel-Cell Hypercar*, Rocky Mountain Institute, Colorado, www.rmi.org/images/other/Trans/T97-09_SpeedingTrans.pdf, accessed 14 January 2005.
33 Fox, J. W. and Cramer, D. R. (1997) *Hypercars: A Market-Oriented Approach to Meeting Lifecycle Environmental Goals*, Rocky Mountain Institute, Colorado, www.rmi.org/images/other/Trans/T97-05_MarketApproach.pdf, accessed 8 June 2005.
34 Fox, J. W. and Cramer, D. R. (1997) *Hypercars: A Market-Oriented Approach to Meeting Lifecycle Environmental Goals*, Rocky Mountain Institute, Colorado, www.rmi.org/images/other/Trans/T97-05_MarketApproach.pdf, accessed 8 June 2005.
35 Cramer, D. R. and Taggart, D. F. (2002) 'Design and manufacture of an affordable advanced-composite automotive body structure', in *Proceedings of the 19th International Battery, Hybrid and Fuel Cell Electric Vehicle Symposium and Exhibition*, www.rmi.org/images/other/Trans/T02-10_DsnManuAdvComp.pdf, accessed 14 January 2007.
36 Lovins, A. B. and Cramer, D. R. (2004) 'Hypercars, hydrogen, and the automotive transition', *International Journal of Vehicle Design*, vol 35, nos 1–2, pp50–85, www.rmi.org/images/other/Trans/T04-01_HypercarH2AutoTrans.pdf, accessed 25 August 2008.
37 Lovins, A. B. and Cramer, D. R. (2004) 'Hypercars, hydrogen, and the automotive transition', *International Journal of Vehicle Design*, vol 35, nos 1–2, pp50–85, www.rmi.org/images/other/Trans/T04-01_HypercarH2AutoTrans.pdf, accessed 25 August 2008.
38 Cramer, D. R. and Taggart, D. F. (2002) 'Design and manufacture of an affordable advanced-composite automotive body structure', in *Proceedings of the 19th International Battery, Hybrid and Fuel Cell Electric Vehicle Symposium and Exhibition*, www.rmi.org/images/other/Trans/T02-10_DsnManuAdvComp.pdf, accessed 14 January 2007.
39 Cramer, D. R. and Taggart, D. F. (2002) 'Design and manufacture of an affordable advanced-composite automotive body structure', in *Proceedings of the 19th International Battery, Hybrid and Fuel Cell Electric Vehicle Symposium and Exhibition*, www.rmi.org/images/other/Trans/T02-10_DsnManuAdvComp.pdf, accessed 14 January 2007.
40 Cramer, D. R. and Taggart, D. F. (2002) 'Design and manufacture of an affordable advanced-composite automotive body structure', in *Proceedings of the 19th International Battery, Hybrid and Fuel Cell Electric Vehicle Symposium and Exhibition*, www.rmi.org/images/other/Trans/T02-10_DsnManuAdvComp.pdf, accessed 14 January 2007.
41 Cramer, D. R. and Taggart, D. F. (2002) 'Design and manufacture of an affordable advanced-composite automotive body structure', in *Proceedings of the 19th International Battery, Hybrid and Fuel Cell Electric Vehicle Symposium and Exhibition*, www.rmi.org/images/other/Trans/T02-10_DsnManuAdvComp.pdf, accessed 14 January 2007.
42 Williams, B. D. et al (1997) *Speeding the Transition: Designing a Fuel-Cell Hypercar*, Rocky Mountain Institute, Colorado, www.rmi.org/images/other/Trans/T97-09_SpeedingTrans.pdf, accessed 14 January 2005.
43 Lovins, A. B. and Cramer, D. R. (2004) 'Hypercars, hydrogen, and the automotive transition', *International Journal of Vehicle Design*, vol 35, nos 1–2, pp50–85, www.rmi.org/images/other/Trans/T04-01_HypercarH2AutoTrans.pdf, accessed 25 August 2008.
44 Cramer, D. R. and Taggart, D. F. (2002) 'Design and manufacture of an affordable advanced-composite

automotive body structure', in *Proceedings of the 19th International Battery, Hybrid and Fuel Cell Electric Vehicle Symposium and Exhibition*, www.rmi.org/images/other/Trans/T02-10_DsnManuAdvComp.pdf, accessed 14 January 2007.

45 Lovins, A. B. and Cramer, D. R. (2004) 'Hypercars, hydrogen, and the automotive transition', *International Journal of Vehicle Design*, vol 35, nos 1–2, pp50–85, www.rmi.org/images/other/Trans/T04-01_HypercarH2AutoTrans.pdf, accessed 25 August 2008.

46 Cramer, D. R. and Taggart, D. F. (2002) 'Design and manufacture of an affordable advanced-composite automotive body structure', in *Proceedings of the 19th International Battery, Hybrid and Fuel Cell Electric Vehicle Symposium and Exhibition*, www.rmi.org/images/other/Trans/T02-10_DsnManuAdvComp.pdf, accessed 14 January 2007.

47 Rose, R. (n.d.) 'Questions and answers about hydrogen and fuel cells', www.fuelcells.org/info/library/QuestionsandAnswers062404.pdf, accessed 19 Oct 2005.

48 Fox, J. W. and Cramer, D. R. (1997) *Hypercars: A Market-Oriented Approach to Meeting Lifecycle Environmental Goals*, Rocky Mountain Institute, Colorado, www.rmi.org/images/other/Trans/T97-05_MarketApproach.pdf, accessed 8 June 2005.

49 Hawken, P., Lovins, A. and Lovins, L. H. (1999) *Natural Capitalism: Creating the Next Industrial Revolution*, Earthscan, London, p35.

50 Cramer, D. R. and Taggart, D. F. (2002) 'Design and manufacture of an affordable advanced-composite automotive body structure', in *Proceedings of the 19th International Battery, Hybrid and Fuel Cell Electric Vehicle Symposium and Exhibition*, www.rmi.org/images/other/Trans/T02-10_DsnManuAdvComp.pdf, accessed 8 June 2005.

51 Fox, J. W. and Cramer, D. R. (1997) *Hypercars: A Market-Oriented Approach to Meeting Lifecycle Environmental Goals*, Rocky Mountain Institute, Colorado, www.rmi.org/images/other/Trans/T97-05_MarketApproach.pdf, accessed 8 June 2005.

52 Wuppertal Institute for Climate, Energy and Environment (2003) *Material Intensity of Materials, Fuels, Transport* (version 2), Wuppertal Institute, Germany, www.wupperinst.org/uploads/tx_wibeitrag/MIT_v2.pdf, accessed 7 May 2008.

53 Lofti, A. (n.d.) 'Automotive recycling', *The Green Pages*, www.lotfi.net/recycle/automotive.html, accessed 19 October 2005.

54 An estimate of hidden costs is given by taking typical values for steel: 8.5kg/kg abiotic material and 75kg/kg water. Wuppertal Institute for Climate, Energy and Environment (2003) *Material Intensity of Materials, Fuels, Transport* (version 2), Wuppertal Institute, Germany, www.wupperinst.org/uploads/tx_wibeitrag/MIT_v2.pdf, accessed 7 May 2008.

55 An estimate of hidden costs is given by taking the mean values for glass, plastic and rubber: 4 kg/kg abiotic material and 284kg/kg water. Wuppertal Institute for Climate, Energy and Environment (2003) *Material Intensity of Materials, Fuels, Transport* (version 2), Wuppertal Institute, Germany, www.wupperinst.org/uploads/tx_wibeitrag/MIT_v2.pdf, accessed 7 May 2008.

56 The assumption of 100 per cent of the Revolution's structure being carbon yields an overestimate in total hidden costs since aluminium, the other material used, has lower hidden costs than carbon fibre. An estimate of hidden costs is given by taking the values for carbon fibre (PAN): 58kg/kg abiotic material and 1795kg/kg water. Wuppertal Institute for Climate, Energy and Environment (2003) *Material Intensity of Materials, Fuels, Transport* (version 2), Wuppertal Institute, Germany, www.wupperinst.org/uploads/tx_wibeitrag/MIT_v2.pdf, accessed 7 May 2008.

8

Worked Example 3 – Electronics and Computer Systems

Significance of electronics and computer design

The world's economy is highly dependent on fast, reliable computers to provide a plethora of information and communications services to governments, businesses and the typical web-surfer. Large banks of computers are linked together and mounted in racks to provide the computing power for companies, but this infrastructure typically comes with large requirements on resources and energy, and produces substantial volumes of waste. Like other engineering systems investigated in this book, computer and electronics systems are traditionally designed with incremental engineering improvement processes, and thus are equally likely to receive Factor 4–10 (75–90 per cent) gains in resource productivity through Whole System Design (WSD). (Note: This worked example will focus on the hardware design of a server, although some related factors of building infrastructure are also briefly discussed.)

Server systems overview

Servers are software applications that carry out tasks on behalf of other software applications called 'clients'. On a network, the server software is run on a computer and acts as the gateway to sharing resources that the client software – which is run on user computers – wants to access (see Figure 8.1 for a simple description of the client–server model). Thus servers must be capable of multitasking – handling interactions with multiple client devices.

In large capacity networks, such as ones servicing many clients or workstations or for internet service providers, multiple server systems are required. Often, up to 42 servers are connected together in a 'rack' and many racks may be required. Consequently, multiple server systems or 'data centres', can occupy whole rooms. Data centres attract overhead costs, including:

- technical staff salaries;
- maintenance costs;
- lighting capital and running costs; and
- air-conditioning with ventilation capital and running costs.

The sensitive operational nature of data centres calls for specialist air-conditioning systems that are required to maintain specific humidity and temperature conditions. Maintenance costs of specialist environmental control systems usually far outweigh the energy costs to run the system itself, and the energy costs for a specialist air-conditioning system are higher than for a conventional system. The overhead costs plus the capital and running costs of the racks and servers themselves make establishing and running a data centre an expensive operation.

Performance characteristics

The following server performance characteristics are critical and usually emphasized during server design, in particular the first three:[1]

1. *Reliability*: Reliable and robust service – server dropouts, even for a short period, can cost a company dearly in lost revenue and potential liability;

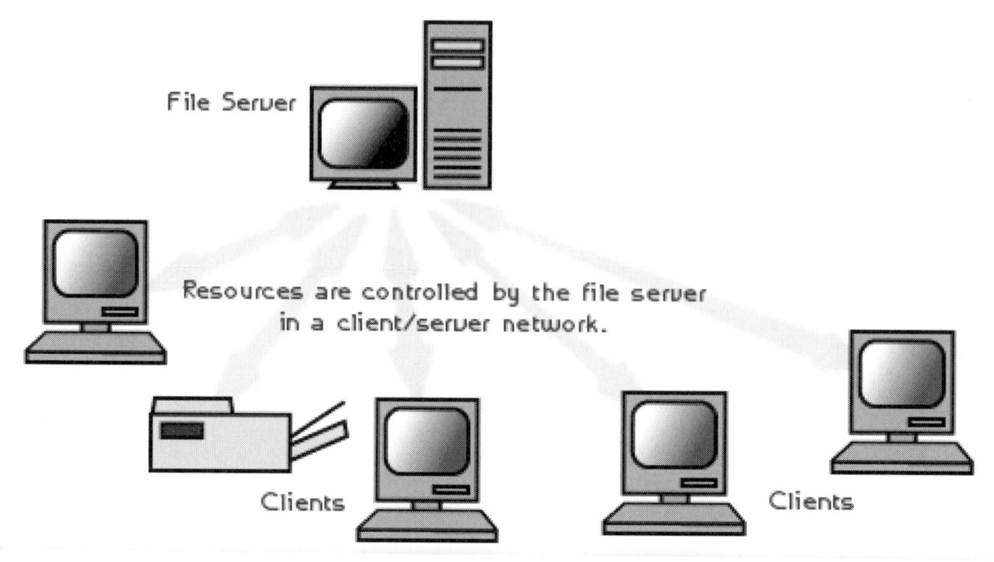

Source: University of South Florida (2005)[2]

Figure 8.1 *Simple diagram of client–server system set-up*

2 *Availability*: 24/7 service and traffic handling – with today's e-business environment, all peripheral devices must be available around the clock, not just during business hours;
3 *Serviceability*: Regular routine servicing – uninterrupted operation is aided by a variety of diagnostic tools and replaceable components;
4 *Scalability*: Achieved by 1) using more central processing units (CPUs) and more powerful input/output (I/O) technology and/or 2) connecting multiple servers;
5 *Manageability*: There are a variety of issues to be addressed – performance monitoring, capacity expansion, system configurability, remote management, automatic or manual load balancing/distribution, and task scheduling; and
6 *Security*: Includes features such as user access control to resources and subsystems, user authentication, intrusion detection, and cryptographic technologies.

There are also operational performance characteristics[3] that are not critical to the average customer but important to the network technician, who is responsible for setting up the server(s) and providing the appropriate climate conditions in the data centre.

The operational performance characteristics influence the running costs of the system, which can be a substantial portion of the total life-cycle cost:

- *Space*: Since multiple server systems occupy whole rooms, they may incur significant rent costs. Conversely, servers are prone to more local heating if tightly packed. At the server level, managing the wiring between servers and peripheral devices is also a challenge that is exacerbated by large, complex servers.
- *Power*: Servers are power-hungry. A server system alone in a data centre can account for a substantial portion of a company's power costs. Supplying a large source of uninterrupted power also requires, at a minimum, a back-up power source.
- *Thermal*: Servers are usually run continuously and thus generate and dissipate more heat than a typical high-speed, high-power desktop personal computer. Most data centres are equipped with air-conditioners to help maintain mild room temperatures so that servers do not overheat and fail. The air-conditioners are relatively high powered and thus contribute substantially to a company's power costs.

Worked example overview

Servers incorporate both server software applications and computer hardware. Servers are one of many subsystems that comprise a data centre. Other subsystems include racks, electrical connections, computer room air-conditioners (CRACs) and the room in the building. Data centres, themselves, are a subsystem of an IT system, which also includes end-user equipment such as computers, printers, faxes, scanners and communications technology, as well as the management of all these technological assets during commissioning, operation, decommissioning and end of life. Further information about the Whole System Design of IT systems is available in The Natural Edge Project's freely available online textbook *Energy Transformed: Sustainable Energy Solutions for Climate Change Mitigation*, 'Lecture 5.3: Opportunities for energy efficiency in the IT industry and services sector';[4] and lecture series *Sustainable IT: Reducing Carbon Footprint and Materials Waste in the IT Environment*.[5] The following worked example focuses on server hardware, although some related factors of building infrastructure are briefly discussed.

Note that, since the development of the following worked example, IT technology has progressed from the conventional solution towards the Whole System Design solution described in this chapter. While the technologies discussed may be outdated, the worked example demonstrates the application of Whole System Design effectively.

Recall the elements of applying a WSD approach discussed in Chapters 4 and 5:

1 Ask the right questions;
2 Benchmark against the optimal system;
3 Design and optimize the whole system;
4 Account for all measurable impacts;
5 Design and optimize subsystems in the right sequence;
6 Design and optimize subsystems to achieve compounding resource savings;
7 Review the system for potential improvements;
8 Model the system;
9 Track technology innovation; and
10 Design to create future options.

The following worked example will demonstrate how the elements can be applied to computer servers using two contrasting examples: a conventional server versus the Hyperserver concept, developed by the Rocky Mountain Institute. The application of an element will be indicated with a shaded box.

Design challenge

Design a server for a data centre comprising 336 servers.

Design process

The following sections of this chapter present:

1 *Conventional design solution*: Conventional system with limited application of the elements of WSD;
2 *WSD solution*: Improved design using the elements of WSD; and
3 *Performance comparison*: Comparison of the economic and environmental costs and benefits.

Conventional computer system design

Select suitable components for the system

Figure 8.2 shows a schematic of a conventional server and indicates the power consumption of the major components. Note that the CPU (70W) and power supply (33W) are the biggest power consumers.

CPU

Conventional servers are designed around high-power, high-speed Central Processing Units (CPUs) that are commonly found in desktop personal computers, such as Intel's Pentium 4 processor, which consumes about 70–75W of power regardless of whether it is running at full capacity or at idle.[6] These processors are selected for their high computational power despite the availability of less power-hungry processors, such as those in laptops and other mobile devices, whose computational power is only slightly less.

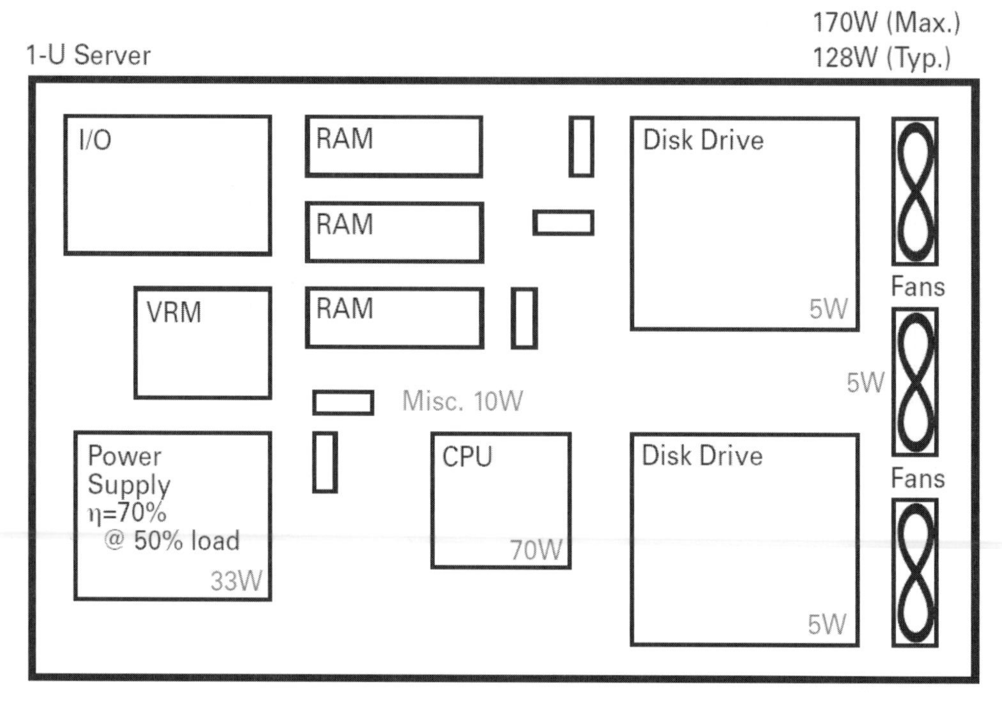

Source: Eubank et al (2003)[7]

Figure 8.2 *Schematic of a conventional server, including power consumption*

Power supply

While power supplies are sized to handle the maximum server load, the server spends most of the time running at 30 to 50 per cent of this.[8] Consequently, a single power supply powering the server would also run at 30 to 50 per cent of its maximum load. However, servers usually incorporate identical, coupled power supplies for redundancy, which run at about 15–30 per cent of their maximum loads.[9] The problem is that power supplies are not designed to run at such low loads and their efficiency drops off rapidly below about 40 per cent load, as shown in Figure 8.3. Running at low load, coupled with other inefficiencies such as multiple current conversions, sees the efficiency of a typical server power supply drop to about 50 per cent or less.[10] This imbalance between maximum load efficiency (which is only 70 to 75 per cent) and low load efficiency arises from the power supply design process, during which power loss is only ever considered at one load – maximum load. This point in the design process is when the heat sink is designed.[11] However, since the power supply usually runs at low load, the heat sink is almost always oversized. Furthermore, power supplies rated greater than 50W, which are common since low efficiency is so predominant, usually require fans for cooling.[12] The fans themselves introduce more heat.

Power supplies are usually coupled with an additional redundant power supply in order to provide an uninterruptable source of power. Typically, the redundant power supplies are always on and operate at low load (20–25 per cent of nameplate) and with low efficiency.[13]

Calculate the cost of the system

Only about half of the power going into a data centre is fed to the servers; the other half is used for overhead energy services such as lighting, air-conditioning and uninterruptable power supplies (UPS). Only about 10 per cent of the air-conditioning power is used to cool at the processor level, while about 50 per cent is used to cool at the data centre level.[14] The higher cooling load at the data centre level is partly due to 'coolth' losses in

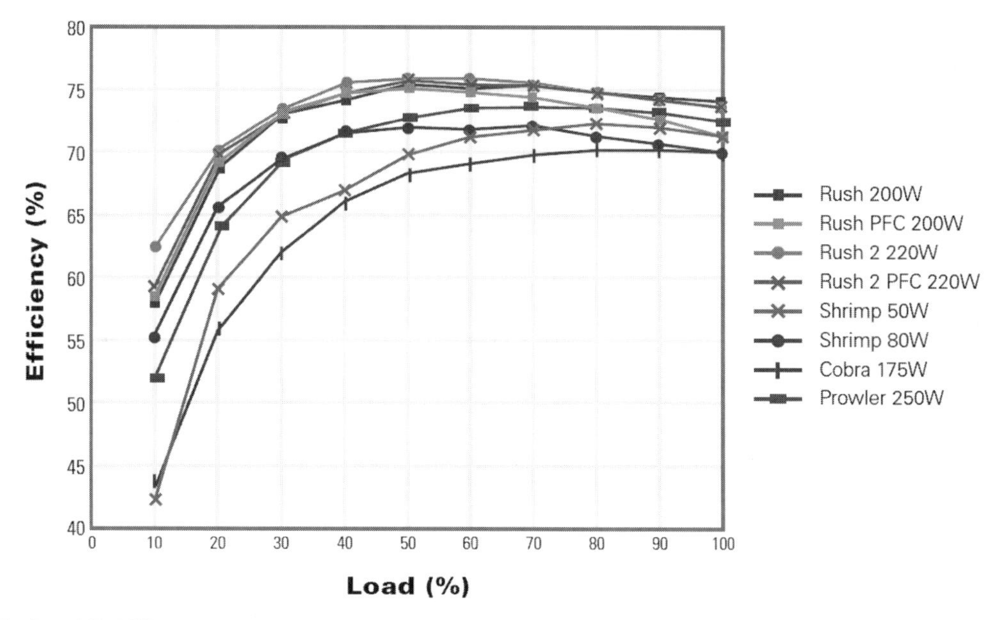

Source: Eubank et al (2003)[15]

Figure 8.3 *Energy efficiencies over full load spectrum of various power supplies*

cooling air when it interacts with hot outgoing air.[16] After incorporating these overhead energy services, the total running cost of a server in a data centre is double what is expected. The ratio of the total power demand of the data centre to the total power demand of the servers is called the 'delivery factor'.[17] For a conventional server data centre the delivery factor is 1.97.[18]

Consider an average conventional server with a three-year life operating in a data centre in Australia, where the cost of electrical power is AU$0.10/kWh (2006 price for large energy users) for a typical large building. The total running cost per watt of power delivered to a conventional server data centre is given by multiplying the cost per kWh by the total running time over the service life of a server by the delivery factor of the data centre:

Running cost/W (AU$0.10/kWh × 0.001kW/W) × (8766 hours/year × 3 years) × (1.97) = AU$5.18/W

For a typical AU$6000, 128W, 0.8A, 12kg server (including external power supply) the total three year running cost is:

Total running cost 128W × AU$5.18/W = AU$663

Now consider a data centre with 336 servers (8 racks of 42 servers):

- The capital cost of the servers is 336 × AU$6000 = AU$2.02 million;
- The running cost of the data centre over 3 years is 336 × AU$663 = AU$222,781;
- The current draw for the servers is 336 × 0.8A = 269A; and
- The mass of the servers is 336 × 12 = 4032kg.

WSD computer system design

Determine a strategy for optimizing all performance characteristics

Element 3: Design and optimize the whole system

The Hyperserver concept[19] was developed using a WSD methodology,[20] and it demonstrates the 60–90 per cent resource productivity improvements that can be made when more emphasis is put on optimizing the *whole system* and factoring the following considerations into design:[21]

Element 1: Ask the right questions

- High energy bills;
- High capital cost;
- Grid dependence;
- Utility distribution charges and delays;
- Risks for the owner/developer;
- Community opposition; and
- Uncaptured opportunities for product sales.

The strategy used to design the Hyperserver is based around reducing the *full resource and economic cost* of the data centre, not just one or two components of it. The power delivered to the server is the key leverage point for reducing energy consumption and costs throughout the whole data centre because the rest of the components (racks, lighting, air-conditioning, ventilation and technical staff) are only there to support the servers. Simpler power-conserving servers mean fewer resources and lower costs for the other components. In other words, reducing the power delivered to the servers will lead to multiple benefits throughout the whole system. Consequently, the strategy used to design the Hyperserver is based around reducing the *full cost of each watt of power delivered to the server*, which means favouring reducing power used continuously over reducing power used intermittently.[22]

Element 7: Review the system for potential improvements

The design strategy is twofold:[23]

1 Reduce or eliminate heat sources:
   - Remove as many energy-intensive components as possible; and
2 Improve heat management:
   - Develop alternative chip-cooling strategies;
   - Optimize heat sinks by choosing the appropriate cooling fin orientation, design and reference values;
   - Remove server box enclosures or minimize the enclosure area to increase airflow; and
   - Put the most heat-intensive and heat-tolerant systems at the top of the rack, where the heat collects (since hot air rises).

Select suitable components for the system

Figure 8.4 shows a schematic of the Hyperserver and indicates the power consumption of the major components. Some components, such as the hard-disk drive and power supply, are external and not shown here. Note that not only is the power consumption of

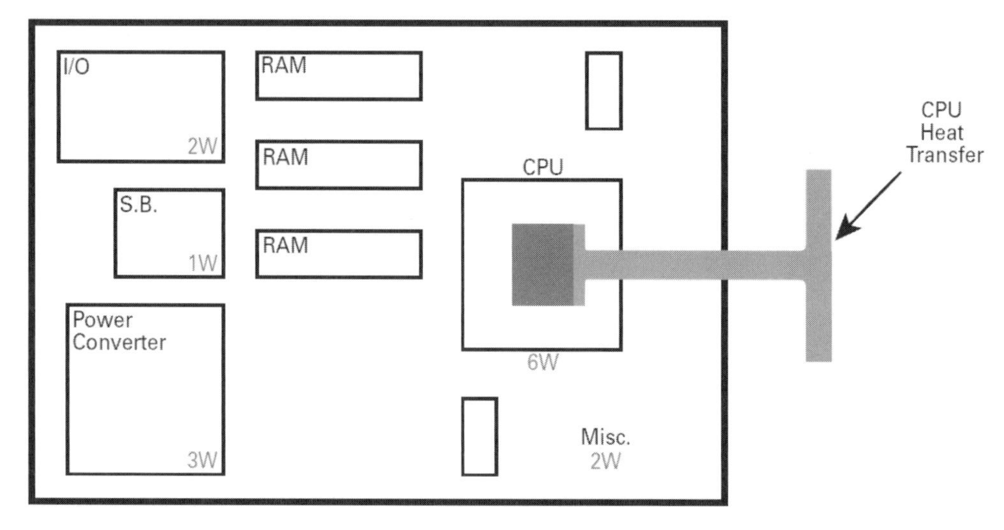

Source: Eubank et al (2003)[24]

Figure 8.4 *Schematic of the WSD server, including power consumption*

the CPU (6W) and power supply (3W) about ten times smaller than in the conventional server, but the power consumption of the other components is also smaller.

Improving heat-intensive components first

Element 5: Design and optimize subsystems in the right sequence

Twenty per cent of the power consumption of servers is due to the fans required to remove heat from heat-intensive components such as the CPU.[25] In the case of the Hyperserver, the design is centred around an ultra-efficient processor, like the Transmetta TM5600, which consumes only 6W of power at load (91 per cent more efficient) and less than 1W (99 per cent more efficient) at idle.[26] As a result, no heat sinks or dedicated fans are required for cooling,[27] making the server much more energy-efficient, smaller and lighter.

External housing of hardware

Hard-disk drives

The hard-disk drives (HDDs) are housed externally in an efficient-to-operate location where the heat generated can be easily removed.[28] Without the space constraint of having to fit the HDD on the server motherboard, larger, shared drives can be used that are more efficient and more reliable than smaller, designated drives.[29] Effectively, only 3W of power is consumed by the drive per server. The operating system, which is usually stored on the drive, is stored on local RAM (DRAM or Flash),[30] and hence more RAM is needed.

Power supply

Like the HDD, the power supply is also housed externally in an efficient-to-operate location.[31, 32] The modularity of the external power supply configuration may favour using a common DC voltage between the power supply and the server, so that all DC to DC conversion can be done by a single 80 per cent efficient power converter that consumes only 3W of power.[33] The external power supply requires an AC to DC power converter, which consumes only 2W of power. The combined 5W power supply consumes 85 per cent less power than the conventional power supply. A common DC voltage to several servers reduces the number of wires, and hence, makes the system easier to handle. Although a common DC voltage configuration can be relatively inefficient,[34] it can be designed such that the inefficiencies are small, as in Appendix 8A at the end of this chapter. The power converter in the server can step up the DC voltage to a variety of voltages such that the components are running at loads that approximate those of peak efficiency.

Element 4: Account for all measurable impacts

The external power supply configuration has multiple benefits:[35]

- Higher efficiency in a direct bus approach;
- Able to supply power at the required capacity rather than at the nameplate rating, which is typically higher;
- A more efficient power supply that can be custom-designed, improved and optimized for life-cycle cost;
- Removal of a major heat source from the board;
- Cheaper equipment: able to use fewer and far more efficient power supplies, both main and redundant;[36]
- More reliable equipment: moves power, fans and heat off-board;
- Quieter: fans removed;
- Size reductions: moving components makes the board smaller.

Redundant power supplies do not need to be on all the time. They can be off or in a low-power standby mode until there is a main power failure.

Server orientation and liquid cooling

Instead of fans, a liquid-based,[37] external cooling system is used, as in Figure 8.5, which shows the traditional method of laying the servers horizontally ('Pizza Boxes') and the alternative – to stand the servers on edge ('Blade Section'). Another possible configuration is to orient the servers diagonally so as to promote natural air convection

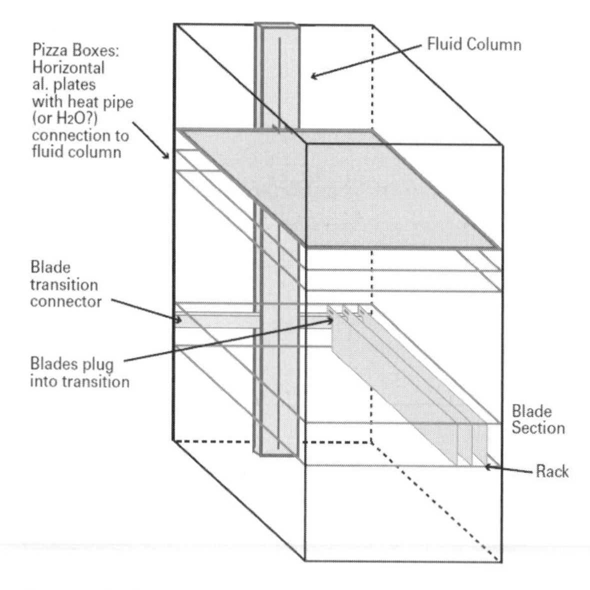

Source: Eubank et al (2003)[38]

Figure 8.5 *Server rack unit with liquid cooling system*

in a zigzag fashion from the bottom of the rack to the top. Vertically and diagonally oriented servers allow air to escape vertically while providing come convective cooling on the way through, whereas horizontally laid servers trap (and stew in) their own hot air. Open- or grated-top racks allow the air to escape from the rack completely.

Element 9:Track technology innovation

The horizontal heat transfer plate is enlarged around the processor – the server's largest single source of heat – and is connected to the fluid column. This configuration has more potential for cooling than the conventional fan systems for two main reasons. Firstly, fluids such as water have a much higher thermal conductivity (0.611W/mK at 27°C)[39] than air (0.0281W/mK at 47°C),[40] which is why cooling fans are not required, hence saving 20–30 per cent of the power consumption.[41] Secondly, the cooling system only requires a small amount of power to circulate the fluid, regardless of how many servers it is cooling, saving a further 25 per cent of power.[42] As a result, the power consumption of a liquid cooling system is only 1W per server.

Air-conditioning

While reducing power consumption in air-conditioning is beyond the scope of this worked example, some low-to-no cost opportunities are mentioned below. Further information about the Whole System Design of air-conditioning systems is available in The Natural Edge Project's freely available online textbook *Energy Transformed: Sustainable Energy Solutions for Climate Change Mitigation*, 'Lecture 2.3: Opportunities for improving the efficiency of HVAC systems'.[43]

- Modern server equipment is more tolerant of temperature and humidity fluctuations than equipment of the past. Consequently, a carefully designed data centre's air-conditioning system does not need to incorporate an elaborate temperature and humidity control. In fact, a conventional split system may suffice. A simple and usually small conventional system saves on expensive maintenance costs, capital costs and power consumption costs.
- In most climates, outdoor air can often be used for passive cooling, especially at night, substantially cutting the daily air-conditioning power consumption. Ambient air can also be used to pre-cool both the air for the air-conditioner and the cooling liquid for the WSD server's liquid cooling system.

Behaviour

While reducing power consumption in data centres through non-technological behavioural means is beyond the scope of this worked example, some no-cost opportunities are mentioned below:

- Often, all lights in a data centre are left on 24/7 when, in fact, they only need to be on for maintenance and upgrades. Turning lights off when the data centre is unoccupied not only saves direct lighting power costs but also indirect air-conditioning costs because the lights themselves also emit heat, which contributes to the cooling load.
- Elsewhere in the building, computers can lead to unnecessary loads on servers even after the building's occupants retire from work for the evening. Computers left on when not in use still make requests to the servers – requests that yield

zero useful result, such as a request for data to run a user's personalized screen saver, which is stored on the shared memory hardware in the building's data centre. Like lighting, computers also contribute to direct and indirect power costs that can be easily avoided.

Calculate the cost of the system

The data centre savings of a Hyperserver-based system are greater than simply those from the lower power consumption. The extra savings arise primarily in the form of the reduced capital costs of overhead equipment such as air-conditioners. For example, Hyperservers require proportionately less cooling assistance *and* smaller cooling equipment than conventional servers, and the UPS can be incorporated with the AC/DC converter rather than being housed separately, making the UPS smaller and able to respond faster to a failure. As a result, the delivery factor for a Hyperserver data centre is 1.36.[44]

Consider a Hyperserver with a three-year life operating in a data centre in Australia, where the cost of electrical power is AU$0.10/kWh (2006 price for large energy users) for a typical large building. The total running cost per watt of power delivered to a Hyperserver data centre is given by multiplying the cost per kWh by the total running time over the service life of a server by the delivery factor of the data centre:

Running cost/W (AU$0.10/kWh × 0.001kW/W) × (8766 hours/year × 3 years) × (1.36)
= AU$3.58/W

For an AU$2500, 21W, 0.13A, 2.4kg Hyperserver[45] (including external power supply) the total three-year running cost is:

Total running cost 21W × AU$3.58/W = AU$75

Now consider a data centre with 336 Hyperservers (8 racks of 42 servers):

- The capital cost of the servers is 336 × AU$2500 = AU$840,000;
- The running cost of the data centre over 3 years is 336 × AU$75 = AU$25,236;
- The current draw for the servers is 336 × 0.13A = 44A; and
- The mass of the servers is 336 × 2.4 = 806.4kg.

Summary: performance comparisons

Server hardware only

The computing power of the Hyperserver (the speed with which the CPU processes instructions) is only about half that of conventional servers. The lower computing power of the Hyperserver is a result of the smaller CPU, which was chosen for its potential to deliver large energy reductions throughout the whole system (conventional CPUs, in some cases twice as powerful, consume 70–80 per cent more energy). Therefore two Hyperservers are needed to match the computing power of a conventional server. Tables 8.1 and 8.2 and Figure 8.6 compare the performance of a conventional server with a twin-Hyperserver system.

Table 8.1 *Power consumption by the major server components*

	Conventional server (W)	Two Hyperservers (W)
CPU	70	12
Hard-disc drive	10	6
Power supply	33	12
Cooling	5	2
Miscellaneous[46]	10	10
Total	*128*	*42*

Table 8.2 *Costs and operating performance comparisons between a conventional server and hyperservers*

	Conventional server	Two Hyperservers	Reduction
Capital Costs (AU$)	6000	5000	17%
Running costs (AU$)	663	150	77%
Power (W)	128	42	67%
Current (A)	0.8	0.26	68%
Mass (kg)	12	4.8	60%

Server hardware plus software control

The performance of the twin-Hyperserver system can be improved further by implementing advanced dynamic resource allocation (DRA). DRA is heavily software-oriented and thus is not discussed in detail here. Briefly, it involves features such as sharing resources and controlling the power feed to resources depending on demand. Advanced DRA can save a further 30 per cent to 50 per cent of power consumption in a data centre,[47] keeping the overall power consumption to 21–30W per twin-Hyperserver system. For a twin-Hyperserver DRA system with power consumption of 28W, the running cost, including data centre energy overhead, is $100 over three years, as indicated in Table 8.3 and Figure 8.6.

Multiple benefits

The Hyperserver outperforms the conventional server in every category given in Tables 8.2 and 8.3. And there are still more benefits to the Whole System server design, as demonstrated by a commercially available server developed by RLX, which shares many of the WSD features with the Hyperserver. The RLX server:

- Requires 1/8 of the space of a conventional server, so the 336 server system fits in a single rack as opposed to 8 racks;
- Has solid-state electrical connections (instead of wires) and redundant power supplies are used, making the overall system more reliable;
- Does not require tools to install additional servers, so expansion of computing abilities is straightforward; and
- Uses 12 times fewer ethernet cables per rack, making management substantially easier.

Progress in industry: Hewlett-Packard

HP Labs, Hewlett-Packard's central research organization, is using a process consistent with WSD. Their whole-system-style 'chip core to cooling tower' approach is used in the design of the 'Smart Chip', 'Smart System' and 'Smart Data Center' – three projects focusing on three levels of computer system design. Together, these three levels of focus form a 'computational continuum', because, as Chandrakant Patel, Director of HP's Sustainable IT Ecosystem Lab,[48] notes, 'if one were to bound this continuum, then one might say that the *data centre is the computer*'. As a result, many of the features of the three projects overlap.

Improving the whole system

HP Labs offer a number of suggestions for optimizing a computer system at the chip,[49] system[50] and data centre levels:

- Cooling resources should be flexible and scalable at all levels.[51] If the cooling mechanism at any level (especially the chip and system levels) is momentarily inadequate, some components could

Table 8.3 *Cost and operating performance comparisons between a conventional server and hyperservers with DRA*

	Conventional server	Two Hyperservers with DRA	Reduction
Capital Cost (AU$)	6000	5000	17%
Running cost (AU$)	663	100	85%
Power (W)	1000	28	84%
Current (A)	0.8	0.17	79%
Mass (kg)	12	4.8	60%

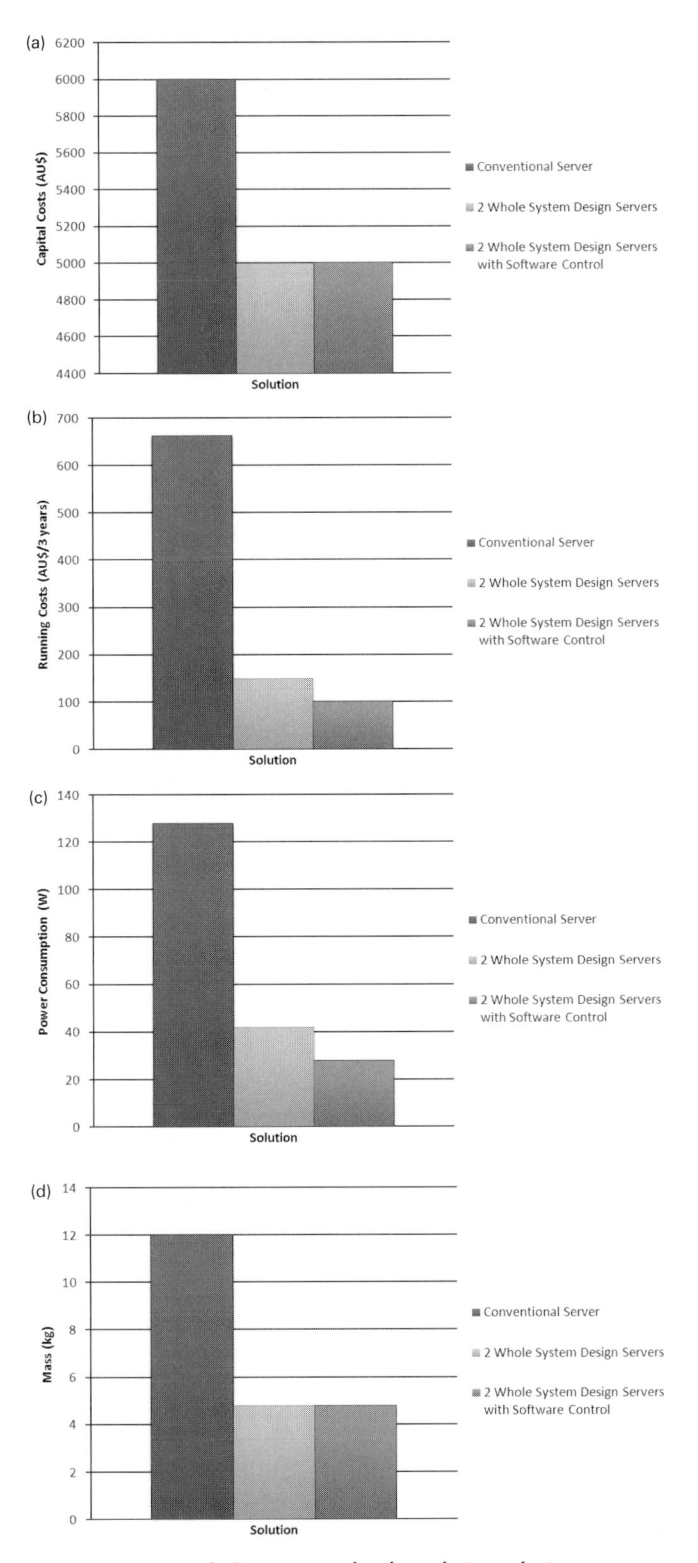

Figure 8.6 *Comparing the three design solutions*

be shut down and the workload could be taken up by another component in a different location.[52]

- *Chip level*:[53] Cooling systems should dynamically adapt to the non-uniform heat distribution on the chip. The current heat-sink-dependent configuration relies on a large temperature difference at the interface between the chip and the heat sink. The large temperature difference results in irreversibilities and destruction of exergy. An example of a more efficient chip-cooling technology involves putting a coolant between the chip and a condenser plate. The right coolant[54] can help dissipate air while minimizing irreversibilities and destruction of exergy.
- *System level*:[55] Heat in a server or rack should be rejected to the surroundings efficiently. Fans should be optimized for the nominal air flow rate and pressure drop across the server or rack, and should be variable-speed. An example of an efficient fan-cooled system involves a single, centralized, variable-speed fan feeding air through many valves and then across channels of components (similar to centralized power supply used in the Hyperserver); each valve controls the air across its own channel. When the components in a channel are at idle, the associated valve should close and the centralized fan's speed should be adjusted.
- *Data centre level*: Hot exhaust air should not mix with cold inlet air.[56] Local cooling at the rack should be favoured over general space cooling.[57] A sensor-actuator control system that incorporates (at a minimum) sensors at the inlet and outlet of the servers should be implemented.[58]
- Decision policies for sensor-actuator control systems should be in combinations of the following strategies:[59]
  - Thermal management-based: Optimize the temperature;
  - Energy efficiency-based: Maximize energy efficiency;
  - Irreversibility-based: Minimize thermodynamic irreversibilities by minimizing mixing of hot and cold air flows;[60]
  - Exergy-based:[61] Minimize the destruction of exergy, including heat dissipated by components; and
  - Performance-based: Optimized computational performance.

Measuring performance

HP Labs suggest that the cost of computing should:

- Be measured by quantities that are applicable at the chip, system and data centre levels;
- Be relevant globally; and
- Provide uniform evaluation.[62]

They suggest measuring cost using MIPS (million instructions per second) per unit of exergy destroyed.[63]

The earliest chips had an efficiency of about 6 MIPS/W and modern chips have an efficiency of about 100 MIPS/W.[64] The MIPS capability of chips will continue to increase further, but will eventually be limited by the high power and cooling requirements that come with extremely high MIPS.[65] At this limit, MIPS per unit exergy is a valuable measurement when comparing chips to determine which configuration holds the most potential for progress.[66]

An equation has been developed for exergy destruction of a modern chip package consisting of all components, from circuit board to heat sink.[67] The equation sets exergy destruction equal to the sum of three terms:[68]

> The first term represents the total exergy destroyed from the electricity supply to the sink base, and is mostly due to the rejection of high-quality electrical energy as low-quality heat. The second term represents the exergy loss due to temperature differences along the fins of the heat sink. The last term indicates the exergy lost due to fluid friction in the airflow. As might be expected, reducing thermal resistance and fluid pressure drop are the two most straightforward ways of lowering exergy consumption in the thermal infrastructure. It is not, however, immediately clear what impact the power consumption of the processor may have on the package exergy loss.

An equation has also been developed for energy destruction in a data centre.[69] Studies on exergy at the data centre level[70] show that there is an optimal computer room air-conditioning (CRAC) air flow rate for minimizing exergy destruction. Exergy destruction increases sharply at flow rates lower or higher than the optimal rate.[71]

Appendix 8A

An issue preventing DC transmission in many applications is that power losses through heat dissipation can be significant. To understand this issue better, consider the following equation:

Power (W) = Voltage (V) × Current (A)

Heat dissipation is correlated with current. A solution is thus simply to transmit the power with high voltage and low current. However, high voltage (or high current) is a safety hazard, especially in the external power supply configuration where the power will be transmitted via a wire. Furthermore, a simple non-isolated DC/DC converter cannot efficiently convert a high voltage to a low voltage, such as is required for the Hyperserver.

A solution is to incorporate an isolated DC/DC converter between the AC/DC converter and the server power supply, as in Figure 8A.1. The isolated DC/DC converter can efficiently convert a high bus voltage to, say, 12V. Simple non-isolated DC/DC converters can then efficiently convert that 12V to the few volts required for each Hyperserver. Safety risks can be minimized by placing the isolated DC/DC converter near the AC/DC converter and thus minimizing the length of the high voltage portion of the power supply system.

Hence the power conversion from mains power to server load can be performed with the following equipment:

- *Mains AC to DC:* A boost power-factor-corrected converter[73] operating at 240V AC input and 400V DC output can achieve 96 per cent conversion efficiency.[74]
- *DC to isolated bus DC:* A soft-switched half-bridge converter can achieve above 93 per cent conversion efficiency.[75]
- *Isolated bus DC to non-isolated server load:* A simple, hard-switching buck converter can step down 12V to 5V or 3.3V at a conversion efficiency of at least 92 per cent.

Thus the total efficiency will be at least 0.96 × 0.93 × 0.92 = 82 per cent.

Cost

Compared to conventional power supply architecture, the architecture in Figure 8A.1 has slightly more power supplies (non-isolated DC/DC converters replace the

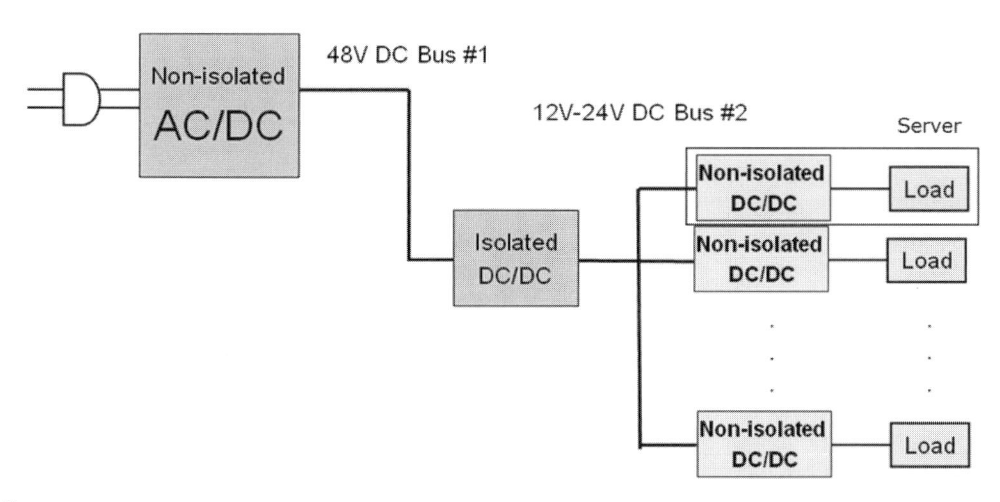

Source: Lu[72]

Figure 8A.1 *Power supply architecture incorporating an intermediate DC–DC conversion to achieve high conversion efficiency*

AC/DC converters, plus a few extra isolated DC/DC converters) but has all of the other benefits listed above in this worked example. The most significant benefit is the overall reduced size and cost of the power supplies due to the high operating efficiency. Furthermore, the architecture in Figure 8A.1 incorporates a centralized AC/DC converter, rather than multiple distributed AC/DC converters. Although the centralized converter needs to be larger in order to handle more power, the total number of controllers, sensors, heat sinks, and plastic and mechanical parts is reduced.

Notes

1 Haghighi, S. (2002) 'Server computer architecture', in J. L. Gaudiot et al, 'Computer architecture and design', in V. Oklobdzija (ed) *Computer Engineering Handbook*, CRC Press, Boca Raton, FL, pp5.4–5.5.
2 Winkelman, R. (2005) *An Educator's Guide to School Networks*, University of South Florida, Chapter 6: 'Software', http://fcit.usf.edu/network/default.htm, accessed 23 March 2008.
3 Haghighi, S. (2002) 'Server computer architecture', in J. L. Gaudiot et al, 'Computer architecture and design', in V. Oklobdzija (ed) *Computer Engineering Handbook*, CRC Press, Boca Raton, FL, p5.5.
4 Smith, M., Hargroves, K., Stasinopoulos, P., Stephens, R., Desha, C. and Hargroves, S. (2007) *Energy Transformed: Sustainable Energy Solutions for Climate Change Mitigation*, The Natural Edge Project, Australia, 'Lecture 5.3: Opportunities for energy efficiency in the IT industry and services sector', www.naturaledge project.net/Sustainable_Energy_Solutions_Portfolio.aspx, accessed 10 April 2008.
5 Stasinopoulos, P., Hargroves, K., Smith, M., Desha, C. and Hargroves, S. (2008) *Sustainable IT: Reducing Carbon Footprint and Materials Waste in the IT Environment,* The Natural Edge Project, Australia, www.naturaledgeproject.net/SustainableIT.aspx, , accessed 10 April 2008.
6 Feng, W. (2002) *The Bladed Beowulf: A Cost-Effective Alternative to Traditional Beowulfs*, IEEE International Conference on Cluster Computing (IEEE Cluster), September 2002, Chicago, IL, http://public.lanl.gov/feng/Bladed-Beowulf.pdf, accessed 23 March 2008.
7 Eubank, H. et al (2003) *Design Recommendations for High-Performance Data Centers*, Rocky Mountain Institute, Snowmass, CO, p34.
8 Eubank, H. et al (2003) *Design Recommendations for High-Performance Data Centers*, Rocky Mountain Institute, Snowmass, CO, p41.
9 Eubank, H. et al (2003) *Design Recommendations for High-Performance Data Centers*, Rocky Mountain Institute, Snowmass, CO, p41.
10 Eubank, H. et al (2003) *Design Recommendations for High-Performance Data Centers*, Rocky Mountain Institute, Snowmass, CO, p41.
11 Eubank, H. et al (2003) *Design Recommendations for High-Performance Data Centers*, Rocky Mountain Institute, Snowmass, CO, p42.
12 Eubank, H. et al (2003) *Design Recommendations for High-Performance Data Centers*, Rocky Mountain Institute, Snowmass, CO, p41.
13 Eubank, H. et al (2003) *Design Recommendations for High-Performance Data Centers*, Rocky Mountain Institute, Snowmass, CO, p39.
14 Shah, A. J. et al (2004) 'An exergy-based control strategy for computer room air-conditioning units in data centers', in *Proceeding of the 2004 ASME International Mechanical Engineering Congress and Exposition*, 13–19 November, Anaheim, CA.
15 Eubank, H. et al (2003) *Design Recommendations for High-Performance Data Centers*, Rocky Mountain Institute, Snowmass, CO, p41.
16 Shah, A. J. et al (2004) 'An exergy-based control strategy for computer room air-conditioning units in data centers', in *Proceeding of the 2004 ASME International Mechanical Engineering Congress and Exposition*, 13-19 November, Anaheim, CA.
17 Eubank, H. et al (2003) *Design Recommendations for High-Performance Data Centers*, Rocky Mountain Institute, Snowmass, CO, p43.
18 Eubank, H. et al (2003) *Design Recommendations for High-Performance Data Centers*, Rocky Mountain Institute, Snowmass, CO, p15.
19 The 'Hyperserver' concept was developed at the Rocky Mountain Institute Design Centre Charrette, 2–5 February 2003.
20 Eubank, H. et al (2003) *Design Recommendations for High-Performance Data Centers*, Rocky Mountain Institute, Snowmass, CO, p11.
21 Eubank, H. et al (2003) *Design Recommendations for High-Performance Data Centers*, Rocky Mountain Institute, Snowmass, CO, p11.
22 Eubank, H. et al (2003) *Design Recommendations for High-Performance Data Centers*, Rocky Mountain Institute, Snowmass, CO, p41.
23 Eubank, H. et al (2003) *Design Recommendations for High-Performance Data Centers*, Rocky Mountain Institute, Snowmass, CO, p37.
24 Eubank, H. et al (2003) *Design Recommendations for High-Performance Data Centers*, Rocky Mountain Institute, Snowmass, CO, p35.
25 Eubank, H. et al (2003) *Design Recommendations for High-Performance Data Centers*, Rocky Mountain Institute, Snowmass, CO, p49.
26 Eubank, H. et al (2003) *Design Recommendations for High-Performance Data Centers*, Rocky Mountain Institute, Snowmass, CO, p37.

27 Eubank, H. et al (2003) *Design Recommendations for High-Performance Data Centers*, Rocky Mountain Institute, Snowmass, CO, p37.
28 Eubank, H. et al (2003) *Design Recommendations for High-Performance Data Centers*, Rocky Mountain Institute, Snowmass, CO, p38.
29 Eubank, H. et al (2003) *Design Recommendations for High-Performance Data Centers*, Rocky Mountain Institute, Snowmass, CO, p38.
30 Eubank, H. et al (2003) *Design Recommendations for High-Performance Data Centers*, Rocky Mountain Institute, Snowmass, CO, p38.
31 Eubank, H. et al (2003) *Design Recommendations for High-Performance Data Centers*, Rocky Mountain Institute, Snowmass, CO, p39.
32 The issues associated with Electromagnetic Interference (EMI) may be magnified when using an externally housed power supply. Any power supply must have at least one stage of EMI filter to comply with Electromagnetic Compatibility (EMC) regulations. For an internally housed power supply, the radiated-EMI can be partially shielded by the server's metal case. However, for an externally housed power supply, a larger EMI filter and/or better shielding may be required. If required, a larger EMI filter or better shielding will increase the cost of the power supply only slightly, especially compared to the overall savings that an externally housed power supply can generate.
33 Eubank, H. et al (2003) *Design Recommendations for High-Performance Data Centers*, Rocky Mountain Institute, Snowmass, CO, p39.
34 Unlike with AC transmission, power losses via DC transmission can be significant. This is a reason why AC transmission is common in conventional server design. Conventional thinking says that, at best, the amount of power dissipated will govern the number of servers that a single common DC voltage can supply, and hence the number of AC/DC converters required. Appendix 8A in this chapter presents an alternative configuration.
35 Eubank, H. et al (2003) *Design Recommendations for High-Performance Data Centers*, Rocky Mountain Institute, Snowmass, CO, p39.
36 Although the chance of having to engage the redundant power supply is reduced, there is still a need to incorporate it to ensure continuous electricity supply to servers. However, since the common DC power supply feeds several servers (centralized power), the total number of main and redundant power supplies is reduced. Also, since the common DC power supply is more efficient, the size and cost of both the main and the redundant power supply are reduced.
37 Patel, C. (2003) 'A vision for energy-aware computing – From chips to data centers', in *Proceedings of the International Symposium on Micro-Mechanical Engineering*, 1–3 December. Patel suggests that liquid cooling is inevitable for processor cooling due to the ineffectiveness of heat sinks at dissipating the large amount of heat generated by increasingly denser chips. Processor cooling is discussed in 'Progress in industry: Hewlett-Packard'. Here we show how liquid cooling can be applied at the rack level.
38 Eubank, H. et al (2003) *Design Recommendations for High-Performance Data Centers*, Rocky Mountain Institute, Snowmass, CO, p49.
39 Mills, A. F. (1999) *Heat Transfer* (second edition), Prentice Hall, Upper Saddle River, NJ, p894.
40 Mills, A. F. (1999) *Heat Transfer* (second edition), Prentice Hall, Upper Saddle River, NJ, p888.
41 Eubank, H. et al (2003) *Design Recommendations for High-Performance Data Centers*, Rocky Mountain Institute, Snowmass, CO, p50.
42 Eubank, H. et al (2003) *Design Recommendations for High-Performance Data Centers*, Rocky Mountain Institute, Snowmass, CO, p49.
43 Smith, M., Hargroves, K., Stasinopoulos, P., Stephens, R., Desha, C. and Hargroves, S. (2007) *Energy Transformed: Sustainable Energy Solutions for Climate Change Mitigation*, The Natural Edge Project, Australia, 'Lecture 2.3: Opportunities for improving the efficiency of HVAC systems', www.naturaledgeproject.net /Sustainable_Energy_Solutions_Portfolio.aspx, accessed 10 April 2008.
44 Eubank, H. et al (2003) *Design Recommendations for High-Performance Data Centers*, Rocky Mountain Institute, Snowmass, CO, p11.
45 Based on RLX's blade servers – among the few WSD servers on the market before 2003, when the Hyperserver was conceived.
46 Miscellaneous power consumed by components such as network interface cards.
47 Eubank, H. et al (2003) *Design Recommendations for High-Performance Data Centers*, Rocky Mountain Institute, Snowmass, CO, p34.
48 Patel, C. et al (2005) 'Smart chip, system and data center enabled by advanced flexible cooling resources', 21st IEEE Semi-Therm Symposium, IEEE CPMT Society, San José, CA, pp78–85.
49 'Chip' is an alternate terminology for microprocessor or processor.
50 The system level refers to the server level or sometimes the rack level.
51 Patel, C. et al (2005) 'Smart chip, system and data center enabled by advanced flexible cooling resources', 21st IEEE Semi-Therm Symposium, IEEE CPMT Society, San José, CA, pp78–85.
52 Patel, C. (2003) 'A vision for energy-aware computing – From chips to data centers', in *Proceedings of the*

International Symposium on Micro-Mechanical Engineering, Tsuchiura and Tsukuba, Japan, 1–3 December.

53 Patel, C. (2003) 'A vision for energy-aware computing – From chips to data centers', in *Proceedings of the International Symposium on Micro-Mechanical Engineering*, Tsuchiura and Tsukuba, Japan, 1–3 December.

54 The 'right' coolant would be the type for which a phase change is reversible. For example, water freezing to ice is a reversible process because the ice can be melted into water again. Boiling an egg, on the other hand, is an irreversible process.

55 Patel, C. (2003) 'A vision for energy-aware computing – From chips to data centers', in *Proceedings of the International Symposium on Micro-Mechanical Engineering*, 1–3 December.

56 Patel, C. (2003) 'A vision for energy-aware computing – From chips to data centers', in *Proceedings of the International Symposium on Micro-Mechanical Engineering*, Tsuchiura and Tsukuba, Japan, 1–3 December.

57 Shah, A. J. et al (2004) 'An exergy-based control strategy for computer room air-conditioning units in data centers', in *Proceedings of the 2004 ASME International Mechanical Engineering Congress and Exposition*, 13–19 November, Anaheim, CA.

58 Patel, C. (2003) 'A vision for energy-aware computing – From chips to data centers', in *Proceedings of the International Symposium on Micro-Mechanical Engineering*, Tsuchiura and Tsukuba, Japan, 1–3 December.

59 Patel, C. et al (2005) 'Smart chip, system and data center enabled by advanced flexible cooling resources', 21st IEEE Semi-Therm Symposium, IEEE CPMT Society, San José, CA, pp78–85.

60 When hot and cold air streams mix, they create a mild temperature air stream from which neither the heat or 'coolth' can ever be recovered again without additional energy being applied; hence there is an 'irreversible' loss of energy.

61 Exergy is the maximum theoretical work attainable when multiple systems at different states interact to equilibrium. Exergy is dependant on the reference environment in which the systems interact. Cited in Moran, M. J. and Shapiro, H. N. (1999) *Fundamentals of Engineering Thermodynamics* (fourth edition), John Wiley and Sons.

62 Patel, C. (2003) 'A vision for energy-aware computing – From chips to data centers', in *Proceedings of the International Symposium on Micro-Mechanical Engineering*, Tsuchiura and Tsukuba, Japan, 1–3 December.

63 Patel, C. (2003) 'A vision for energy-aware computing – From chips to data centers', in *Proceedings of the International Symposium on Micro-Mechanical Engineering*, Tsuchiura and Tsukuba, Japan, 1–3 December.

64 Shah, A. et al (2005) 'Impact of chip power dissipation on thermodynamic performance', 21st IEEE Semi-Therm Symposium, IEEE CPMT Society, San José, CA, pp99–108.

65 Shah, A. et al (2005) 'Impact of chip power dissipation on thermodynamic performance', 21st IEEE Semi-Therm Symposium, IEEE CPMT Society, San José, CA, pp99–108.

66 Shah, A. et al (2005) 'Impact of chip power dissipation on thermodynamic performance', 21st IEEE Semi-Therm Symposium, IEEE CPMT Society, San José, CA, pp99–108.

67 Shah, A. et al (2005) 'Impact of chip power dissipation on thermodynamic performance', 21st IEEE Semi-Therm Symposium, IEEE CPMT Society, San José, CA, pp99–108.

68 Shah, A. et al (2005) 'Impact of chip power dissipation on thermodynamic performance', 21st IEEE Semi-Therm Symposium, IEEE CPMT Society, San José, CA, pp99–108.

69 Shah, A. J. et al (2004) 'An exergy-based control strategy for computer room air-conditioning units in data centers', in *Proceedings of the 2004 ASME International Mechanical Engineering Congress and Exposition*, 13–19 November, Anaheim, CA.

70 Shah, A. J. et al (2004) 'An exergy-based control strategy for computer room air-conditioning units in data centers', in *Proceedings of the 2004 ASME International Mechanical Engineering Congress and Exposition*, 13–19 November, Anaheim, CA.

71 Shah, A. J. et al (2004) 'An exergy-based control strategy for computer room air-conditioning units in data centers', in *Proceedings of the 2004 ASME International Mechanical Engineering Congress and Exposition*, 13–19 November, Anaheim, CA.

72 Dylan Lu, University of Sydney, personal communication in June 2006.

73 Regulation from IEC 6-1000-3-2 states that every power supply that has input power greater than 75W needs to limit the input current harmonics using a power-factor-corrected AC/DC converter.

74 Spiazzi, G. et al (2003) 'Performance evaluation of a Schottky SiC power diode in a Boost PFC Application', *IEEE Transactions on Power Electronics*, vol 18, no 6, pp1249–1253.

75 Korotkov, S. et al (1997) 'Soft-switched asymmetrical half-bridge DC/DC converter: Steady-state analysis. An analysis of switching processes', Telescon 97: The Second International Telecommunications Energy Special Conference, 22–24 April, pp177–184, http://ieeexplore.ieee.org/xpl/RecentCon.jsp?punumber =5266, accessed 28 March 2007.

9

Worked Example 4 – Temperature Control of Buildings

Significance of low-energy homes

Today even the most expensive and luxurious homes and modern buildings can generate electricity on-site through renewable energy technologies such as rooftop photovoltaic cells, solar thermal collectors and small-scale wind turbines. These technologies can reduce the building's demand for electricity from the grid and hence save money. In some countries, any excess electricity can be sold back to the grid, which is particularly profitable during peak load periods.

However, before options for renewable power generation are considered, the amount of electricity used by the home needs to be reduced. Many homes are very inefficient users of electricity and there are many cost-effective options for reducing demand without reducing comfort or service.[1] Making the building a low-energy consumer will then reduce the infrastructure needed to generate the electricity on-site, and allow more of the electricity produced to then be sold back to the grid.

Lowering electricity consumption in a home can be done in a number of ways, such as through the choice of energy-efficient appliances (in 2006 General Electric launched its eco-efficient range of home appliances as part of the Ecomagination programme) and through energy-saving practices and behaviour; however, the need for providing a comfortable temperature in the home is one of the largest electricity consumers. The demand for electricity for air-conditioning in the summer and heating in the winter for homes and commercial buildings is driving increased peak electricity demand, which then drives the need to build new power plants and maintain distribution infrastructure, the poles and cables. This phenomenon was discussed in detail in Chapter 4 of The Natural Edge Project's *The Natural Advantage of Nations* publication. Hence, for many reasons reducing the need for air-conditioning and making it less energy-intensive is a very important mechanism for reducing demand for electricity.

Minimizing heat transfer into or out of the home (heat loss in winter and heat gain in summer) can be done in a number of ways. The goal is to control the internal temperature of the home within a preferred temperature range. The temperature inside the home is usually increased by people, electrical appliances and lighting; however, the main need for temperature control in homes is driven by the outside conditions.

In many cases the design of a home does not look holistically for ways to reduce the need for temperature control but rather compensates with a reliance on heating, ventilation and air-conditioning (HVAC) systems. However, there are a range of strategies for reducing the heat load and extremes of temperature. Firstly, high temperatures during summer can be reduced by blocking solar radiation with trees, generous overhangs, awnings and verandas, and through situating water structures or planted trellises upwind of the home. Secondly, much can be done to regulate the air temperature in a house through passive solar design: orientation of the house, thermal mass, natural ventilation and insulation. There is a range of very cost-effective ways to improve insulation, such as topping up ceiling insulation, cavity wall insulation, pelmets and appropriate curtains, double glazing windows, weather seals for doors, draught stoppers under doors, roof vents, ensuring windows are sealed, and insulating around the hot water system.

Such investments can significantly reduce the need for regular use of ceiling fans or a HVAC system and allow a home to be adequately heated or cooled quickly with relatively small and cheap fans, air-conditioners and heaters.[2] Placing manually adjustable windows to take advantage of prevailing winds and thermal convection to ventilate living spaces can similarly reduce the use and size of HVAC systems.

HVAC systems themselves can be designed in different ways. The Australian Greenhouse Office has a training module on HVAC systems.[3] The US Department of Energy provides an overview of the wide range of ways to achieve temperature control in buildings.[4] It is important also to consider when designing such systems how energy-efficient other air-conditioning and heating systems are as well.[5]

This design worked example will look at the home as a system and consider a range of options and their economic impacts.

Worked example overview

HVAC systems are one of many subsystems that comprise a building's temperature control system, as discussed above. Most of these subsystems are applicable to the design of all types of buildings, including houses, commercial buildings and industrial workshops. Further information about the Whole System Design of commercial buildings and HVAC systems is available in The Natural Edge Project's freely available online textbook *Energy Transformed: Sustainable Energy Solutions for Climate Change Mitigation*, 'Lecture 2.2: Opportunities for energy efficiency in commercial buildings' and 'Lecture 2.3: Opportunities for improving the efficiency of HVAC systems'.[6]

The following worked example focuses on cooling systems in houses. Specifically, it focuses on working through the cooling load calculations required for sizing a residential HVAC system. These calculations can be quite complex and calculations for commercial buildings are even more so. For this reason, this chapter presents only the major components of the calculations required for a very simple house. The analysis is sufficient in demonstrating the effect of Whole System Design (WSD) on building design.

In the Australian building industry, software packages such as second generation NatHERS[7] programs (AccuRate, FirstRate5[8] and BERSPro[9]) and other rating tools[10] are often used. These software packages streamline the modelling and calculation process, but care is required in ensuring that the published protocols for use are adhered to.

The cooling load calculations in this worked example will be based on the 'Cooling Load Temperature Difference/Cooling Load Factor (CLTD/CLF)' method – a simplified, hand-calculation method presented by the American Society of Heating, Refrigerating and Air-Conditioning Engineers (ASHRAE) in their 1997 Handbook.[11] The later versions of this publication (2001 onwards – the book is updated usually every four years) present another method known as the 'Radiant Time Series' method, which gives an exact solution, but requires computer-aided numerical computation. In this worked example, northern hemisphere data from the ASHRAE 1997 Handbook are used for southern hemisphere calculations, under the assumption that asymmetries between the two hemispheres are negligible.

Recall the elements of applying a WSD approach discussed in Chapters 4 and 5:

1. Ask the right questions;
2. Benchmark against the optimal system;
3. Design and optimize the whole system;
4. Account for all measurable impacts;
5. Design and optimize subsystems in the right sequence;
6. Design and optimize subsystems to achieve compounding resource savings;
7. Review the system for potential improvements;
8. Model the system;
9. Track technology innovation; and
10. Design to create future options.

The following worked example will demonstrate how the elements can be applied to temperature control in domestic buildings using two contrasting examples: a conventional house versus a WSD house. The main focus is on Elements 2, 3, 4 and 5. The application of the other elements will be indicated with a shaded box.

Design challenge

Design a north-facing, single-storey house for a family of two parents and three children. The house is located in Canberra, Australia. Include a HVAC system if necessary.

Design process

The following sections of this chapter present:

1. *General solution*: A solution for any single-storey house in Canberra, Australia, incorporating a set of assumptions;
2. *Conventional design solution*: Conventional system with limited application of the elements of WSD;
3. *WSD solution*: Improved system using the elements of WSD; and
4. *Performance comparison*: Comparison of the economic and environmental costs and benefits.

General Solution

The house is in a temperate climate where it is hot in summer and cold in winter. Consequently, the house may require both cooling and heating equipment.[12] However, for the purposes of this worked example, only the summer scenario will be considered. The winter scenario is covered in the ASHRAE Handbook.[13]

Climate

The aim of the worked example is to design a house that can maintain the following interior climate conditions:

- Design temperature: 24°C; and
- Design humidity ratio: 50 per cent

The house is located in Canberra, Australia (Latitude: 35.15°S), where the summer climate has the following characteristics:[14]

- Dry bulb temperature (0.4 per cent[15]): 32.5°C;
- Daily range of dry bulb temperature: 13.3°C; and
- Humidity ratio (0.4 per cent): 13.7 per cent.

Since the dry bulb temperature in Canberra (32.5°C) is substantially greater than the design temperature (24°C), a HVAC system will probably be required.

Assumptions

The following assumptions are made for this worked example:

- The house is approximated as a single large room, so heat gain from interior, unconditioned space, q_p, is not applicable;
- Daylight penetration affects all areas of the building evenly;[16]
- Average cloudiness and other weather conditions;
- Steady state heat transfer conditions;
- The effects of moisture are ignored;
- The outside and interior temperatures are uniformly distributed;[17]
- There is no exterior artificial lighting;
- Thirty per cent of all lamps are on during times of maximum cooling load;
- All lamps operate for 2000 hours per year;
- Ignore the cost of any structural resizing;
- Electrical appliances release an average of 25 per cent of input power as heat;[18]
- Fifty per cent of all electrical appliances are on during times of maximum cooling load;
- Electrical appliances consume power at an average rate of 10 per cent of their input power at all times; ignore installation costs of components;
- The occupants of any solution behave similarly;[19] and
- The assumptions and conditions of any equations and tables used in the references are relevant and accurate.

Calculating cooling load

Element 8: Model the system

The size of the HVAC system is determined by the maximum cooling load, which is determined by the heat gain. Heat gain comes from two types of sources, external and internal, each of which contributes various cooling loads. The cooling load equations are given below.[20] Table 9.1 describes each symbol in the equations.

External

Heat gain through opaque surfaces (walls, roof and doors):

$$q_W = A_W U_W (CLTD) \text{ for each exterior wall}$$

$$q_R = A_R U_R (CLTD)$$

$$q_D = A_D U_D (CLTD) \text{ for each door}$$

Heat gain through translucent surfaces (windows and skylights):

$$q_G = A_G (GLF) + A_{GS} (GLF)_S \text{ for each window and skylight}$$ [21]

Heat gain due to infiltration of outside air through leaks:

$$q_{IS} = 1.2 Q \Delta t$$

$$q_{IL} = 3 Q \Delta w,$$

where:

$$Q = \frac{1000(ARC)(RV)}{3600}$$

Internal

Heat gain due to occupants:

$$q_{OS} = H_{OS} (CLF)_O \text{ for each occupant}$$

$$q_{OL} = H_{OL} \text{ for each occupant}$$

Heat gain due to artificial lighting:

$$q_L = P_L F_U F_A (CLF)_L (1 - e_L) \text{ for each lamp}$$ [22]

Heat gain due to electrical appliances:

$$q_A = H_A (CLF)_A \text{ for each appliance}$$ [23]

Heat gain from interior, unconditioned space (ignored):

$$q_p = A_P U_P \Delta t_p \text{ for each interior partition or wall}$$

Table 9.1 *Symbol nomenclature for design cooling load equations*

Symbol	Description	Units
A_D	Area of door	m^2
A_G	Unshaded area of window or skylight	m^2
A_{GS}	Shaded area of window or skylight	m^2
A_P	Area of interior partition or wall	m^2
A_R	Area of roof projected onto horizontal (excluding skylights)	m^2
A_W	Area of exterior wall (excluding windows)	m^2
ACH	Air exchange rate (air exchanges per hour)	1/hr
CLF_A	Cooling load factor for electrical appliances (time delay factor)	
CLF_L	Cooling load factor for artificial lighting (time delay factor)	
CLF_O	Cooling load factor for occupants (time delay factor)	
$CLTD$	Cooling load temperature difference for each surface	K or °C
e_L	Energy efficiency of lamp	
F_U	Usage factor (fraction of time in use)	
F_A	Allowance factor (for additional losses)	
GLF	Glass load factor for each unshaded window or skylight	W/m^2
GLF_S	Glass load factor for each shaded window or skylight	W/m^2
H_A	Heat gain due to electrical appliances	W
H_{OL}	Latent heat gain due to occupants	W
H_{OS}	Sensible heat gain due to occupants	W
L_P	Duct losses	
q_A	Cooling load for electrical appliances	W
q_C	Maximum cooling load	W
q_D	Cooling load for doors	W
q_{DES}	Design cooling load	W
q_G	Cooling load for windows and skylights	W
q_{IL}	Latent cooling load for infiltration	W
q_{IS}	Sensible cooling load for infiltration	W
q_L	Cooling load for artificial lighting	W
q_{OL}	Latent cooling load for occupants	W
q_{Os}	Sensible cooling load for occupants	W
q_P	Cooling load for interior, unconditioned space	W
q_R	Cooling load for roof	W
q_W	Cooling load for walls	W
RV	Room volume (in this case, volume of house)	m^3
Δt	Temperature differential between outside and interior	K or °C
Δt_P	Temperatures differential across an interior partition or wall	K or °C
U_D	Overall heat transfer coefficient of door	W/m^2K
U_P	Overall heat transfer coefficient of interior partition or wall	W/m^2K
U_R	Overall heat transfer coefficient of roof	W/m^2K
U_W	Overall heat transfer coefficient of wall	W/m^2K
Δw	Humidity ratio differential between outside and interior	kg/kg

Source: ASHRAE (1997)[24]

The maximum cooling load, q_C, is simply the sum of all individual cooling loads. The design cooling load, q_{DES}, accounts for duct losses.

$$q_{DES} = q_C(1+L_D)$$

Conventional design solution

Design the building and determine the occupancy characteristics

For simplicity, the house has a simple rectangular floor plan. The north-facing frontage is narrow to allow room for a double carport on the western side of the house and a small walkway on the eastern side. The conventional house has the following features:

- Northern façade: 10m wide × 3m high; two windows: 2m × 2m each; no shade; door: 1m × 2m;
- Eastern façade: 20m wide × 3m high; four windows: 1.5m × 1m each; 40 per cent tree shade;
- Southern façade: 10m wide × 3m high; two windows: 2m × 2m each; no shade (no direct sun); door: 1m × 2m;
- Western façade: 20m wide × 3m high; four windows: 1.5m × 1m each; 40 per cent tree shade;
- Roof/attic/ceiling: pitched; foil and fibre glass bat insulation equivalent to R3.5; small overhang;
- Exterior walls: fibre glass bat insulation equivalent to R1.5;
- Windows: single glazing, 3.2mm glass, sliding, wood frame, draperies;
- Doors: solid core flush door;
- Loose construction (many gaps in the building envelope that enable air drafts);
- Interior electrical appliances: total input power of 4kW;[25] and
- Interior electrical lighting: 30 100W incandescent lamps.

The occupants are expected to be performing the following activities during times of maximum cooling load:

- Man: light work;
- Woman: moderately active office work;
- Child 1: playing (equivalent to light bench work);
- Child 2: playing (equivalent to light bench work); and
- Child 3: seated reading (equivalent to seated in a theatre).

Calculate the design cooling load

The values in Table 9.2 yield the results in Table 9.3, which are compared graphically in Figure 9.1. Note that major sources of heat gain are *external* heat gains.

Calculate the cost of the system

Costs are presented for house components that differ from the WSD solution.

HVAC

The HVAC system needs to be rated at least q_{DES} = 8.5kW.

Select Fujitsu ART30LUAK/AOT30LMBDL (ducted):[26]

Output power: 8.8kW

Input power: 3.3 kW

Capital cost[27] = AU$2500

In Canberra a residential air-conditioner runs for about 150h/yr.[28] With electricity costing about AU$0.18/kWh (2006 price for residential supply), the running cost of the HVAC system in the conventional solution is:

Running cost = (3.3kW)(150h/yr)(AU$0.18/kWh) = AU$89/yr.

Windows

There is 28m^2 of single glazing and 72m of wooden window frame.

Single glazing: AU$400/m^2

Wooden window frames with seals: AU$20/m of window perimeter

Capital cost = (28m^2)($400/m^2) + (72m)(AU$20/m) = AU$12,640

Table 9.2 *Values used to calculate design cooling load, Q_{DES}, for the house*

Symbol	Value	Source
A_D	(2m)(1m) = 2m^2	Given
A_G	north, south: (2m)(2m)(2) = 8m^2 east, west: (0.6)(1.5m)(1m)(4) = 3.6m^2 roof: 0m^2 (no skylight)	Given
A_{GS}	north, south: 0m^2 east, west: (0.4)(1.5m)(1m)(4) = 2.4m^2 roof: 0m^2 (no skylight)	Given
A_P	n/a	
A_R	(20m)(10m) = 200m^2	Given
A_W	north, south: (10m)(3m) – A_G – A_D = 20m^2 east, west: (20m)(3m) – A_G – A_D = 54m^2	Given
ACH	0.7 per hour	ASHRAE Handbook, Table 8, p27.4
CLF_A	1	HVAC not operating 24hr/day
CLF_L	1	ASHRAE Handbook, p28.52[29]
CLF_O	1	ASHRAE Handbook, p28.52[30]
$CLTD$	north: 6K, south: 4K; east, west: 10K roof: 23K	ASHRAE Handbook, Table 1, p27.2
e_L	0.15	APS[31]
F_U	0.3	Given
F_A	1	ASHRAE Handbook, p28.8[32]
GLF	North:[33] 71.5W/m^2, south: 60W/m^2 east, west: 145W/m^2	ASHRAE Handbook, Table 3, p. 27.3
GLF_S	north, east, south, west: 60W/m^2	ASHRAE Handbook, p27.5[34]
H_A	(0.25)(0.5)(4kW)(1000W/kW) = 500W	Given
H_{OL}	55W + (0.85)(55W) + (0.75)(140W) + (0.75)(140W) + (0.75)(30W) – 307.5W	ASHRAE Handbook, Table 3, p28.8
H_{OS}	75W + (0.85)(75W) + (0.75)(80) + (0.75)(80) + (0.75)(65W) = 334.25W	ASHRAE Handbook, Table 3, p28.8
L_D	0.08	ASHRAE Handbook, p27.6[35]
RV	(20m)(10m)(3m) = 600m^3	Given
Δt	32.5°C-24°C = 8.5°C	Outside t: ASHRAE Handbook, Table 5, p26.27, dry bulb temp. for Canberra
Δt_P	n/a	
U_D	2.21W/m^{2}K	ASHRAE Handbook, Table 6, p24.13
U_P	n/a	
U_R	1/3.5 = 0.286W/m^{2}K	[36]R3.5 roof insulation: McGee (2005)[37] for Canberra
U_G	5.05W/m^{2}K	ASHRAE Handbook, Table 5, p29.8
U_W	1/1.5 = 0.667W/m^{2}K	R1.5 wall insulation: McGee (2005)[38] for Canberra
Δw	0.5–0.137 = 0.363	Interior w: ASHRAE Handbook, Table 5, p26.27 for Canberra

Electrical appliances

The interior electrical appliances will change for the WSD solution but their capital cost will be assumed to be unchanged. The input power of all electrical appliances totals 4kW and all appliances run 10 per cent of the time on average.

With electricity costing about AU$0.18/kWh (2006 price for residential supply), the running cost of the electrical appliances in the conventional solution is:

Running cost = (0.1)(4kW)(24hr/day)(365days/yr)(AU$0.18/kWh) = AU$631/yr

Electrical lighting

All 30 100W lamps operate for 2000 hours per year.

100W lamp: AU$1/lamp for 1000 hour life

Capital cost = (30 lamps)(AU$1/lamp) = AU$30

Table 9.3 *Breakdown of design cooling load components for the conventional solution*

External
q_W = 853W q_R = 1314W q_D = 44W q_G = 2384W q_{IS} = 1190W q_{IL} = 127W
Internal
q_{O_s} = 308W q_{OL} = 334W q_L = 765W q_A = 500W
Totals
q_C = 7820W q_{DES} = 8445W

With electricity costing AU$0.18/kWh (2006 price for residential supply), the running cost of the electrical lighting in the conventional solution is:

Running cost: (30 lamps)(0.1kW)
(2000h/yr)(AU$0.18/kWh) = AU$1080/yr

See Table 9.6 for a summary of costs.

WSD solution

Redesign the building with less heat gain

Element 7: Review the system for potential improvements

In Table 9.3, the design cooling load for the conventional design is dominated by the *external* heat gains, of which the three largest are:

1. q_R: heat gain through the roof;
2. q_G: heat gain through the windows; and
3. q_{IS}: sensible heat gain due to infiltration of outside air through leaks.

Therefore reducing these heat gains offers the greatest potential for reducing the overall design cooling load. That is not to say that the other components should be ignored. In fact, by designing the system as a whole (rather than component-wise), the benefits of any design feature will impact on the operations of other components. A key consideration for maximizing benefits is embodied in Element 5: 'Design and optimize subsystems in the right sequence'. Here, the

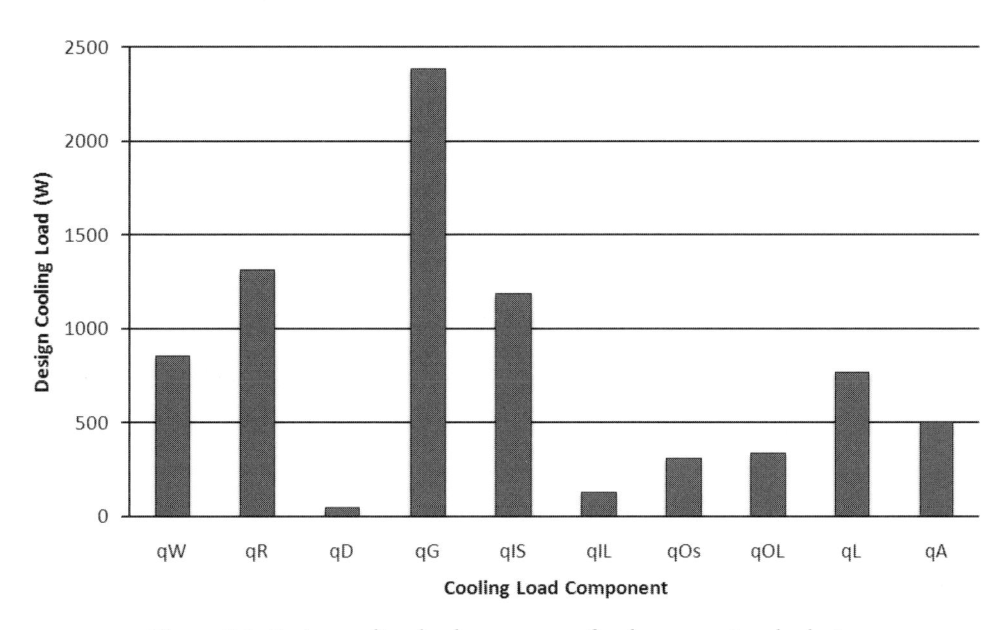

Figure 9.1 *Design cooling load components for the conventional solution*

design and optimization activities that provide multiple benefits are usually prioritized. Determining the right sequence in which to implement the activities requires deeper consideration.

Element 5: Design and optimize subsystems in the right sequence

Passive technologies (like shading devices) reduce the demand for active technologies (like air-conditioning units) as they do not require energy and therefore should be designed and sized first. Each passive technology can both improve the effectiveness of other passive technologies and reduce the overall demand for active technologies. Active technologies, on the other hand, require energy input, which comes at a cost. Optimizing passive and active technologies in the sequence indicated in Figure 9.2 will maximize their effectiveness and the house's overall performance.

The WSD solution for the design challenge of this house considers aspects of building envelope (including orientation, window shading, window types, building materials, landscaping and construction quality); daylighting; artificial (electrical) lighting; electrical appliance selection; and HVAC. Passive solar heating and passive cooling/ventilation are not a focus in this chapter.

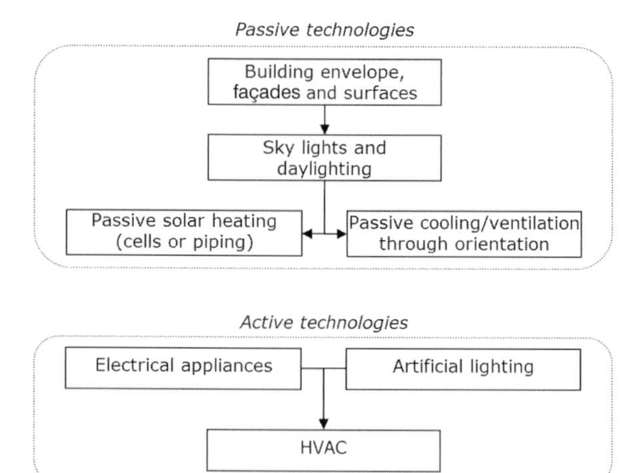

Figure 9.2 *Building feature design sequence for minimizing energy consumption*

Building envelope and daylighting

The house's building envelope can potentially reduce the cooling loads due to the three dominant components of heat gain, q_R, q_G and q_{IS}. Reducing the external dimensions of the whole house will reduce the cooling load substantially, since either area, A, or room volume, RV, appears in the equation for every component of external cooling load. However, for comparability, the external dimensions of the house are maintained.

Since the rectangular floor plan dimensions are maintained, the horizontal area of the roof, AR, is also maintained. Consequently, the first dominating component of heat gain, q_R, is unaffected by changes in building envelope. There are some changes that can reduce q_R by other means. For example, the roof's (and wall's) colour can influence the cooling load temperature difference (CLTD) for the roof. Lighter colours reduce CLTD and hence heat gain through the roof, q_R. The effect of colour on residential buildings is relatively small and is thus ignored. However, the effect of colour on commercial buildings can be significant, potentially reducing the equivalent outside roof temperature from up to 45°C above ambient to within 10°C.[39]

The second dominating component of heat gain, q_G, can be reduced through the combination of four changes:

1 Rotate the house by 90 degrees. The double carport is replaced with a single carport and the side walkway is eliminated so that the north-facing frontage can be made wider. The depth of the house is correspondingly smaller.
2 Have larger windows on the northern and southern façades (and smaller windows on the eastern and western façades) to encourage milder daylighting throughout the day, rather than more intense early and late sun through eastern and western windows. The house's shallower depth improves the relative penetration of natural daylight.
3 Increase the length of the roof overhang and introduce trees on the northern side to provide shade. Since the meridian altitude ('elevation') of the sun is higher during summer than in winter, the roof overhang can be designed such that the amount of direct solar contact on windows is reduced (increasing the shaded area

of the window) in summer while still allowing sufficient passive heating in winter. In this solution shading is assumed to be 40 per cent. In practice, the actual amount of shade provided by an overhang during summer can be calculated using Table 6 in ASHRAE.[40] In a temperate climate, deciduous trees are best since they provide shade in summer and allow solar gain in winter (when it is needed).

4 Use more efficient windows. In this worked example the use of triple glazing with air gaps is explored. The gas gap (usually air or argon) reduces the overall heat transfer coefficient through the windows, U_G, since the gas has a lower thermal conductivity than the glass. An alternative or addition to multiple glazed windows are spectrally selective windows and films, which block most infrared (heat containing) and ultraviolet light, while transmitting a good portion of visible light. Two such commercially available technologies are available from Viracon[41] and V-KOOL.[42]

The third dominating component of heat gain, q_{IS}, can be reduced through 'tighter' construction, which involves both more careful selection of seal-like components and more careful workmanship during construction.

The WSD house has the following building envelope and construction features:

- Northern façade: 20m wide × 3m high; four windows: 2m × 1.5m; 40 per cent roof overhang and tree shade; door: 1m × 2m;
- Eastern façade: 10m wide × 3m high; four windows: 1m × 0.5m; 40 per cent tree shade;
- Southern façade: 20m wide × 3m high; four windows: 2m × 1.5m; no shade;[43] door: 1m × 2m;
- Western façade: 10m wide × 3m high; four windows: 1m × 0.5m; 40 per cent roof overhang and tree shade;
- Roof/attic/ceiling: pitched; foil and fibre glass bat insulation equivalent to R3.5; large overhang contributing to greater window shading on northern façade;
- Exterior walls: fibre glass bat insulation equivalent to R1.5;
- Windows: triple glazing, 12.7mm airspace, sliding, wood frame, draperies;
- Doors: solid core flush door; and
- Medium to tight construction (many gaps in building envelope that enable air drafts).

Passive solar heating

In this worked example, passive solar heating, which is a significant factor in winter heating load, is not a focus because only the summer scenario is being considered. Note that in temperate climates, where winter heating is often required, determining heating load is critical for overall systems optimization, because heat load is dependent on many of the same house components (such as insulation) as is cooling load. For this reason, some of the changes made for the WSD solution were only modest, for example, specifying that the roof overhang would contribute to a percentage of window shading, rather than specifying the actual length of the overhang, which can also influence the amount of passive heat gain in winter. Furthermore, some obvious changes were not made at all. These changes include upgrading the insulation to prevent more heat gain in summer,[44] which would also prevent the desirable day-time heat gain during winter, and reducing the window area on the southern façade to prevent undesirable heat loss during winter, which would also reduce the penetration of daylight. In fact, it is likely that a relatively small window area on the southern façade would be optimal since q_G is several times greater than q_W. However, for comparability, the total window area of the house is maintained at $28m^2$.

Passive cooling/ventilation

In this worked example, passive cooling/ventilation is not considered, because its effects are not significant during the period of maximum cooling load. Since maximum cooling load usually occurs around the hottest part of the day, the house is usually closed up to prevent hot air from leaking in and contributing to heat gain. Note, however, that passive ventilation can play an important role in evening cooling on hot days. The 'CLTD/CLF' method for residential homes used in this worked example does not explicitly account for delay effects.

Electrical appliances

Element 6: Design and optimize subsystems to achieve compounding resource savings

Often, the heat gain from electrical appliances can be reduced at no cost by simply 'shopping around' for the right-sized appliance. Compared to oversized appliances, which sometimes are selected because the customer incorrectly perceives these appliances to be better value for money, the right-sized appliances have several benefits. For example, they are often cheaper to purchase, because they are a lower-capacity appliance; they run near their design loads, which makes them roughly twice as efficient as an appliance running at low load; they emit less heat; and they are generally smaller, lighter and safer to handle. The WSD solution takes into account appliance right-sizing by conservatively assuming that all appliances are 10 per cent more efficient, on average.

The WSD house has the following total input power from electrical appliances:

> Interior electrical appliances: total input power of 3.6kW[45]

Artificial lighting

Artificial electrical lighting is through compact fluorescent lamps. Although fluorescent lamps are about nine times more expensive than incandescent lamps, they have several advantages that make them better value for money. For example, fluorescent lamps last about eight times longer than incandescent lamps; they expend only 30 per cent (as opposed to 85 per cent in the incandescent lamps) of their input energy as heat; and they convert about 70 per cent (as opposed to 15 per cent) of their input energy to light.[46] To overcome the concerns about the quality of light from fluorescent lamps, a number of manufacturers are developing a range of shades of bulb to deliver more-natural coloured interior light.

Compared to the conventional solution, the window configuration of the WSD solution (larger windows on northern and southern façades) allows more natural daylight into the house, especially during the middle of the day when maximum cooling load occurs. Consequently, fewer than the assumed 30 per cent of all lamps would be on during times of maximum cooling load. However, for simplicity, this assumption is maintained as per the conventional solution.

The WSD house has the following artificial lighting equipment:

> Interior electrical lighting: 30 15W compact fluorescent lamps

Occupants

For comparability, occupants are expected to be performing the same activities as in the conventional solution. That is:

- Man: light work;
- Woman: moderately active office work;
- Child 1: playing (equivalent to light bench work);
- Child 2: playing (equivalent to light bench work); and
- Child 3: seated reading (equivalent to seated in a theatre).

Element 3: Design and optimize the whole system

This WSD solution is not the optimal available solution, it is merely an improved solution. Optimizing the solution requires comparing system performance and costs resulting from incorporating all other reasonable combinations of the house components, and then selecting the best-fitting solution for the occupant. Note that simply integrating the best technologies for each component often does not yield an optimal solution,[47] so it is important to consider any house component change against the performance of the whole system, not just the component itself. The design challenge, even with its assumptions and simplifications, allows for several house component changes. Such a task is complex and repetitive and beyond the scope of this chapter.

Calculate the design cooling load

The values in Table 9.4 yield the results in Table 9.5, which are compared graphically in Figure 9.3. Note that the major sources of heat gain are still *external* heat gains.

Table 9.4 *Values used to calculate the design cooling load, Q_{DES}, for the house*

Symbol	Value	Source
A_D	(2m)(1m) = 2m²	Given
A_G	north, south: (0.6)(2m)(1.5m)(4) = 7.2m² east, west: (0.6)(1m)(0.5m)(4) = 1.2m² roof: 0m² (no skylight)	Given
A_{GS}	north, south: (0.3)(2m)(1.5m)(4) = 4.8m² east, west: (0.4)(1m)(0.5m)(4) = 0.8m² roof: 0m² (no skylight)	Given
A_P	n/a	
A_R	(20m)(10m) = 200m²	Given
A_W	north, south: (20m)(3m) – A_G – A_D = 46m² east, west: (10m)(3m) – A_G – A_D = 28m²	Given
ACH	0.4 per hour	ASHRAE Handbook, Table 8, p27.4
CLF_A	1	HVAC not operating 24 hr/day
CLF_L	1	ASHRAE Handbook, p28.52[48]
CLF_O	1	ASHRAE Handbook, p28.52[49]
$CLTD$	north: 6K, south: 4K; east, west: 10K roof: 23K	ASHRAE Handbook, Table 1, p27.2
e_L	0.7	
F_U	0.3	Given
F_A	1.2	ASHRAE Handbook, p. 28.8[50]
GLF	North:[51] 62W/m², south: 50W/m² east, west: 123W/m²	ASHRAE Handbook, Table 3, p27.3
GLF_S	north, east, south, west: 50W/m²	ASHRAE Handbook, p27.5[52]
H_A	(0.25)(0.5)(3.6kW)(1000W/kW) = 450W	Given
H_{OL}	55W + (0.85)(55W) + (0.75)(140W) + (0.75)(140W) + (0.75)(30W) = 307.5W	ASHRAE Handbook, Table 3, p28.8
H_{OS}	75W + (0.85)(75W) + (0.75)(80) + (0.75)(80) + (0.75)(65W) = 334.25W	ASHRAE Handbook, Table 3, p28.8
L_D	0.08	ASHRAE Handbook, p27.6[53]
RV	(20m)(10m)(3m) = 600m³	Given
Δt	32.5°C – 24°C = 8.5°C	Outside t: ASHRAE Handbook, Table 5, p26.27, dry bulb temp. for Canberra
Δt_P	n/a	
U_D	2.21W/m²K	ASHRAE Handbook, Table 6, p24.13
U_P	n/a	
U_R	1/3.5 = 0.286W/m²K	[54]R3.5 roof insulation: McGee (2005)[55] for Canberra
U_G	2.19W/m²K	ASHRAE Handbook, Table 5, p29.8
U_W	1/1.5 = 0.667W/m²K	R1.5 wall insulation: McGee (2005)[56] for Canberra
Δw	0.5 – 0.137 = 0.363	Interior w: ASHRAE Handbook, Table 5, p26.27 for Canberra

Calculate the cost of the system

Costs are presented for house components that differ from the conventional solution.

HVAC

The HVAC system needs to be rated at least q_{DES} = 6.1kW. Select:

Panasonic CS-F24DD1E5/CU-L24DBE5[57]

Output power: 6.3kW

Input power: 2.09kW

Capital cost = [58]AU$1780

In Canberra, a residential air-conditioner runs for about 150h/yr.[59] With electricity costing about AU$0.18/kWh (2006 price for residential supply), the running cost of the HVAC system in the WSD solution is:

Table 9.5 *Breakdown of design cooling load components for WSD solution*

External
q_W = 680W q_R = 1314W q_D = 44W q_G = 1660W q_{IS} = 680W q_{IL} = 73W
Internal
q_{O_s} = 308W q_{OL} = 334W q_L = 41W q_A = 450W
Totals
q_C = 5583W q_{DES} = 6030W

Running cost = (2.09kW)(150h/yr)
(AU$0.18/kWh) = AU$56/yr

Windows

There is 28m^2 of triple glazing and 80m of wooden window frame.

Triple glazing: AU$450/m^2

Wooden window frames with seals:
AU$20/m of window perimeter

Capital cost = (28m^2)(AU$450/m^2)
+ (80m)(AU$20/m) = AU$14,200

Electrical appliances

Appliances in the WSD solution are assumed to be 10 per cent more energy efficient, on average, than those in the conventional solution. It is assumed that the efficient appliances were identified through 'shopping around' and thus their total capital costs are the same as for the less efficient appliances.

The input power of all electrical appliances totals 3.6kW, and all appliances run 10 per cent of the time, on average.

With electricity costing about AU$0.18/kWh (2006 price for residential supply), the running cost of the electrical appliances in the WSD solution is:

Running cost: (0.1)(3.6kW)(24hr/day)
(365 days/yr)(AU$0.18/kWh)
= AU$568/yr

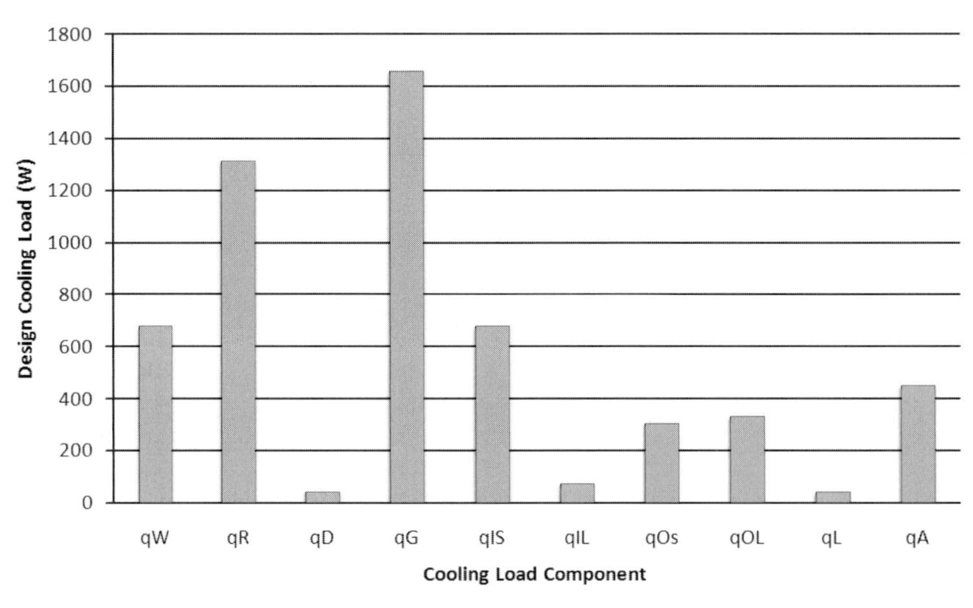

Figure 9.3 *Design cooling load components for the WSD solution*

Electrical lighting

Compared to the conventional solution, the window configuration of the WSD solution (larger windows on northern and southern façades) allows more natural daylight into the house throughout the day. Consequently, all lamps are likely to be on for less than 2000 hours per year. However, for simplicity, the assumption is maintained that all 30 15W lamps operate for 2000 hours per year. The costs of the ballasts for fluorescent lamps can be avoided by selecting the compact variety, which are directly interchangeable with incandescent lamps.

15W lamp: AU$9/lamp for 8000 hour life

Capital cost = (30 lamps)(AU$9/lamp) = AU$270

With electricity costing AU$0.18/kWh (2006 price for residential supply), the running cost of the electrical lighting in the WSD solution is:

Running cost = (30 lamps)(0.015kW)
(2000hr/yr)(AU$0.18/kWh) = AU$162/yr

Additional costs

The WSD solution incorporates some components that do not appear in the conventional solution.

Trees: 4 trees for shading the northern façade
at a cost of AU$100 each

Capital cost = (4 trees)(AU$100/tree) = AU$400

Roof overhang: larger roof overhang to
provide shading at a cost of AU$10/m
of house perimeter

Capital cost = [60](20m + 10m + 20m
+ 10m)(AU$10/m) = AU$600

Construction: Extra seal components for
windows, doors, joints and other locations
of potential air leak at a cost of AU$5/m
of house perimeter

Capital cost = (20m + 10m + 20m
+ 10m)(AU$5/m) = AU$300

See Table 9.6 for a summary of costs.

Summary: performance comparisons

A comparison of system performance and costs helps to highlight the efficacy of WSD for residential housing and the building industry in general.

Cooling load

Table 9.6 and Figure 9.4 compare the cooling loads of the two solutions.

The design cooling load, q_{DES}, for the WSD solution is 29 per cent lower than that of the conventional solution. The majority of the performance improvements are from external sources – 20 per cent lower q_W, 30 per cent lower q_G and 42 per cent lower q_{IL}. The internal cooling loads were similar or the same, because those loads are usually occupant-dependent. The only exception is the artificial lighting, for which the use of fluorescent lamps reduced the associated cooling load by 95 per cent.

Cost

Table 9.7 compares the costs of the two solutions.

The total capital costs for both solutions are dominated by the costs of the windows. However, the results in Table 9.7 only indicate the costs of those house components that differ between the two solutions. The capital costs for the rest of the house (worth hundreds of thousands of dollars) would dwarf the capital costs in Table 9.7. Thus a meaningful comparison considers the *absolute* difference (not percentage difference) in total capital cost. By the same argument, a meaningful comparison of running cost considers the absolute difference in running cost.

With the exception of the fluorescent lamps, it is unlikely that the more expensive components of the WSD solution would be cost-effective on their own. However, when *combined*, the more expensive house components offset some of the total capital cost by making suitable a smaller and cheaper HVAC system. That is, from an economic perspective, the *system* of components is more valuable than the simple sum of individual components.

The main economic advantage of the WSD solution arises in its lower running costs. Table 9.7 shows that for a roughly AU$3000 higher capital cost, the WSD solution saves about AU$1000 per year, which gives a

Table 9.6 *Comparing breakdown of the cooling loads for the two solutions*

Cooling Load Component	Conventional Solution	WSD Solution
	External	
q_W	853W	680W
q_R	1314W	1314W
q_D	44 W	44 W
q_G	2384W	1660W
q_{IS}	1190W	680W
q_{IL}	127W	73W
	Internal	
q_{Os}	308W	308W
q_{OL}	334W	334W
q_L	765W	41W
q_A	500W	450W
	Totals	
q_C	7820W	5583W
q_{DES}	8445W	6030W

payback period of about three years. More importantly, the savings in running cost equate to about $11,000 over 15 years or $15,000 over 30 years (assuming that the cost of electricity remains constant).

Multiple benefits

Element 4: Account for all measurable impacts

WSD/low-energy buildings not only have lower running and long term costs than their conventional equivalents, they also have several other benefits.

- Lower greenhouse gas emissions, since the electricity saved usually comes from electricity producers, of which the largest portion are coal-fired.
- Lower energy consumption, which makes viable renewable energy technologies that currently cost more per unit of power generated than purchasing grid electricity. The capital cost savings of a WSD solution can offset the capital cost of the renewable energy technology, which will thereafter save on electricity costs and eventually pay itself off. On-site renewable energy technologies also improve power service reliability both locally and to the wider community.
- 'Sell or lease faster, and retain tenants better, because they combine superior amenity and comfort with lower operating costs and more competitive terms.

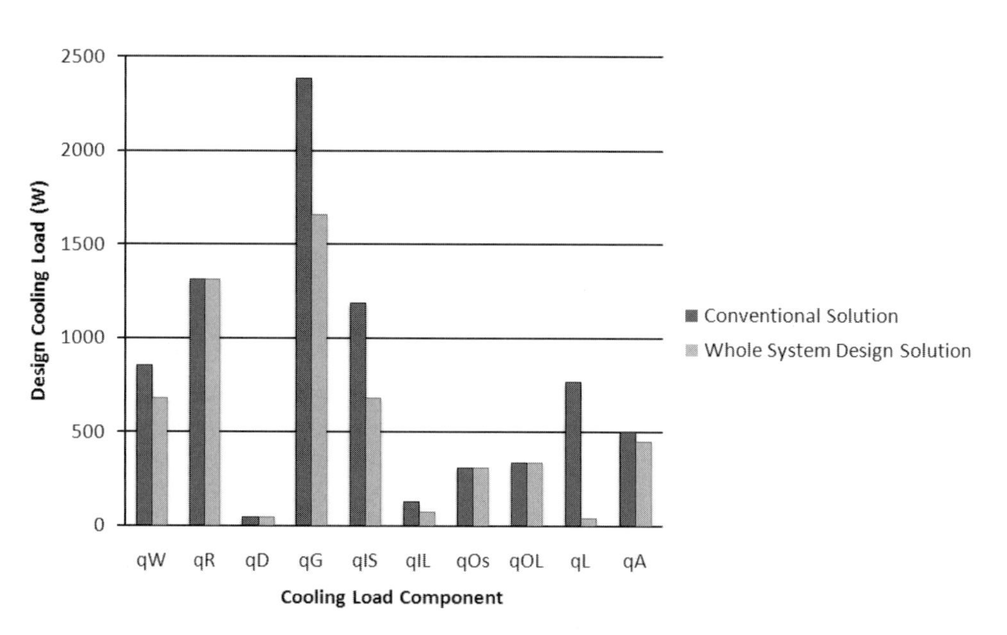

Figure 9.4 *Comparing the cooling loads for the two solutions*

Table 9.7 *Comparing the costs of the two solutions*

House Component	Conventional Solution	WSD Solution
	Capital Cost	
HVAC	AU$2500	AU$1780
Electrical lighting	AU$30	AU$270
Windows	AU$12,080	AU$14,200
Trees		AU$400
Roof overhang		AU$600
Construction		AU$300
Total	AU$14,610	AU$17,550
	Annual Running Cost	
HVAC	AU$89	AU$56
Electrical lighting	AU$1080	AU$162
Electrical appliances	AU$631	AU$568
Total	AU$1800	AU$786
	Life-Cycle Running Cost[61]	
15 years	AU$19,282	AU$8420
30 years	AU$26,577	AU$11,605

The resulting gains in occupancies, rents and residuals all enhance financial returns.'[62]

- Provide 'greater visual, thermal and acoustic comfort, [which] creates a low-stress, high-performance environment that yields valuable gains in labour productivity, retail sales, and manufacturing quality and output. These improvements in turn create a key competitive advantage, and hence further improve real estate value and market performance'.[63] In many organizations, labour costs are several times greater than energy costs. This leverage point turns small labour productivity improvements into large economic savings.
- Require fewer materials to build and operate, because active technologies are smaller or sometimes even eliminated. Reducing active technologies reduces the required structural integrity, noise insulation and heat insulation of the building.

Notes

1 For further information, refer to pages such as Australian Government, 'Your home: Design for lifestyle and the future' at www.yourhome.gov.au/, accessed 7 May 2008, or ACT Home Energy, 'Advisory team fact sheets' at www.heat.net.au/topics.html, accessed 14 April 2007. An overview of the most energy-efficient approaches to buildings and appliances is available at www.gotoreviews.com/archives/metaefficient/, accessed 14 April 2007.

2 See Alternative Technology Association at http://ata.org.au/more-info?page_id=58, accessed 14 April 2007.

3 Australian Greenhouse Office, 'HVAC Training Module' at www.greenhouse.gov.au/lgmodules/wep/hvac/index.html, accessed 14 April 2007.

4 US Department of Energy, 'HVAC links' at www.b4ubuild.com/links/hvac.shtml, accessed 14 April 2007.

5 MetaEfficient provides an overview of efficient appliances and devices at www.metaefficient.com, accessed 14 April 2007.

6 Smith, M., Hargroves, K., Stasinopoulos, P., Stephens, R., Desha, C. and Hargroves, S. (2007) *Energy Transformed: Sustainable Energy Solutions for Climate Change Mitigation*, The Natural Edge Project, Australia, 'Lecture 2.2: Opportunities for energy efficiency in commercial buildings', and 'Lecture 2.3: Opportunities for improving the efficiency of HVAC systems', www.naturaledgeproject.net/Sustainable_Energy_Solutions_Portfolio.aspx, accessed 10 April 2008.

7 See NatHERS (Nationwide House Energy Rating Scheme) website at www.nathers.gov.au/, accessed 7 May 2008.

8 See Sustainability Victoria, 'FirstRate' at www.sustainability.vic.gov.au/www/html/1491-energy-rating-with-firstrate.asp, accessed 7 May 2008.

9 See Solar Logic, 'BERS Pro' at www.solarlogic.com.au/bers-pro, accessed 7 May 2008.

10 See Australian Government, 'Your home: Design for lifestyle and the future' at www.yourhome.gov.au/, accessed 7 May 2008.

11 American Society of Heating, Refrigerating and Air-Conditioning Engineers, Inc. (ASHRAE) (1997) *ASHRAE Handbook: Fundamentals*, ASHRAE, Atlanta.

12 According to the Sustainable Energy Development Office, in Canberra the average annual operating hours for HVAC systems is 150hr/ yr for cooling and 500hr/yr for heating.

13 ASHRAE (1997) *ASHRAE Handbook: Fundamentals*, ASHRAE, Atlanta.

14 ASHRAE (1997) *ASHRAE Handbook: Fundamentals*, ASHRAE, Atlanta, pp26–27.

15 Indicates that the temperature reported is exceeded about 0.4 per cent of the time.

16 This assumption is included for simplicity. The depth to which daylight penetrates a building has a bearing on the demand for interior artificial lighting, in this case electrical lamps. Daylight also introduces heat into a building, so the parts of the building that are daylit are also warmer.

17 This assumption is included for simplicity; however, uniform interior temperatures are not common for several

reasons: 1) building features such as interior walls impede heat transfer; 2) solar heat gain though fenestrations such as windows, doors and skylights results in higher local temperatures; and 3) active (and sometimes passive) temperature control is not usually available in non-main rooms such as the bathroom or the laundry.

18 Actual heat gains for specific appliances are given in ASHRAE (1997) *ASHRAE Handbook: Fundamentals*, ASHRAE, Atlanta, Tables 4, 6, 7, 8, 9A and 9B, pp28.10–28.14.

19 Occupant behaviour plays a large role in building consumption. Energy-aware occupants take more care in turning off electrical appliances and lights when not in use, thus reducing their electricity requirement and running costs while extending the life of the appliances and lights. Energy-aware occupants may also tolerate less comfortable interior conditions (in summer, higher temperature and humidity), thus further reducing their electricity requirements and running costs while extending the life of the HVAC system.

20 ASHRAE (1997) *ASHRAE Handbook: Fundamentals*, ASHRAE, Atlanta, pp27.5, 27.6, 27.13, 28.40.

21 Modified to explicitly incorporate shading effects.

22 Modified to incorporate energy efficiency.

23 Simplified.

24 ASHRAE (1997) *ASHRAE Handbook: Fundamentals*, ASHRAE, Atlanta.

25 Estimate, since the actual appliances used are dependent on the occupants.

26 See Energy Rating at http://search.energyrating.gov.au/air_srch.asp, accessed 21 July 2006.

27 Estimate, since prices for residential HVAC vary substantially.

28 Sustainable Energy Development Office (n.d.) *Your Guide to Energy-Smart Air-Conditioners*, SEDO, Australia, www1.sedo.energy.wa.gov.au/uploads/Air-Conditioners_65.pdf, accessed 7 February 2006.

29 'If the cooling system operates only during occupied hours, the CLFL should be considered 1.0...' ASHRAE (1997) *ASHRAE Handbook: Fundamentals*, ASHRAE, Atlanta, p28.52.

30 'If the space temperature is not maintained constant during the 24h period, a CLF of 1.0 should be used.' ASHRAE (1997) *ASHRAE Handbook: Fundamentals*, ASHRAE, Atlanta, p28.52.

31 'About 15 per cent of the energy [incandescent lamps] use comes out as light – the rest becomes heat.' APS (n.d.) 'Different types of lighting', APS, US, p1, www.aps.com/main/services/business/WaysToSave/BusWaystoSave_9.html, accessed 2 February 2006.

32 'For fluorescent fixtures and/or fixtures that are either ventilated or installed so that only part of their heat goes to the conditioned space.' ASHRAE (1997) *ASHRAE Handbook: Fundamentals*, ASHRAE, Atlanta, p28.8.

33 Corrected as per note (b) in ASHRAE (1997) *ASHRAE Handbook: Fundamentals*, ASHRAE, Atlanta, p27.3.

34 'Glass shaded by overhangs is treated as north glass.' ASHRAE (1997) *ASHRAE Handbook: Fundamentals*, ASHRAE, Atlanta, p27.5. In this chapter's southern hemisphere worked example, shaded glass is treated as south glass.

35 'If all ducts are in the attic space, a duct loss of 10 per cent space sensible cooling load is reasonable.' ASHRAE (1997) *ASHRAE Handbook: Fundamentals*, ASHRAE, Atlanta, p27.6. LP is a loss factor over both sensible and latent cooling loads. To compensate for inaccuracies from including latent cooling load, LP is reduced from 0.1 to 0.08.

36 U=1/R.

37 McGee, C., Mosher, M. and Clarke, D. (2005) 'Insulation: Overview', in *Technical Manual: Design for Lifestyle and the Future* (third edition), Commonwealth of Australia, http://greenhouse.gov.au/yourhome/technical/fs16a.htm, accessed 21 July 2006.

38 McGee, C., Mosher, M. and Clarke, D. (2005) 'Insulation: Overview', in *Technical Manual: Design for Lifestyle and the Future* (third edition), Commonwealth of Australia, http://greenhouse.gov.au/yourhome/technical/fs16a.htm, accessed 21 July 2006.

39 Merritt, F. S. and Ricketts, J. T. (2001) *Building Design and Construction Handbook* (sixth edition), McGraw-Hill, p13.38.

40 ASHRAE (1997) *ASHRAE Handbook: Fundamentals*, ASHRAE, Atlanta, Table 6, p27.4.

41 Viracon (www.viracon.com) have developed SuperwindowTM technology that can reject up to 98 per cent of infrared light.

42 V-KOOL (www.v-kool.com) have developed polyester films that can reject up to 94 per cent of infrared light.

43 The southern façade does not receive direct sunlight, so shading devices do not have an effect on the calculations (GLF and GLF_S for the southern façade are the same). However, installing trees may cool the local air – a factor not accounted for by the calculations.

44 High R insulation can also prevent desirable heat transfer from the interior to the exterior during summer nights.

45 Equal to 10 per cent less than the total input power for electrical appliance in the conventional solution.

46 Department of the Environment, Water, Heritage and the Arts (2008) *Phase-Out of Inefficient Light Bulbs*, Commonwealth of Australia, www.greenhouse.gov.au/energy/cfls/index.html, accessed 15 February 2008; Energy Star (n.d.) *Compact Fluorescent Light Bulbs*, US Environmental Protection Agency and US Department

of Energy, www.energystar.gov/index.cfm?c=cfls.pr_cfls, accessed 15 February 2008.

47 Hawken, P, Lovins, A. and Lovins, H. (1999) *Natural Capitalism: Creating the Next Industrial Revolution*, Earthscan, London, p117.

48 'If the cooling system operates only during occupied hours, the CLFL should be considered 1.0.' ASHRAE (1997) *ASHRAE Handbook: Fundamentals*, ASHRAE, Atlanta, p28.52.

49 'If the space temperature is not maintained constant during the 24h period, a CLF of 1.0 should be used.' ASHRAE (1997) *ASHRAE Handbook: Fundamentals*, ASHRAE, Atlanta, p28.52.

50 'Recommended value of 1.20 for general applications.' ASHRAE (1997) *ASHRAE Handbook: Fundamentals*, ASHRAE, Atlanta, p28.8.

51 Corrected as per note (b) in ASHRAE (1997) *ASHRAE Handbook: Fundamentals*, ASHRAE, Atlanta, p27.3.

52 'Glass shaded by overhangs is treated as north glass.' ASHRAE (1997) *ASHRAE Handbook: Fundamentals*, ASHRAE, Atlanta, p27.5. In this chapter's southern hemisphere worked example, shaded glass is treated as south glass.

53 'If all ducts are in the attic space, a duct loss of 10 per cent space sensible cooling load is reasonable.' ASHRAE (1997) *ASHRAE Handbook: Fundamentals*, ASHRAE, Atlanta, p27.6. LP is a loss factor over both sensible and latent cooling loads. To compensate for inaccuracies from including latent cooling load, LP is reduced from 0.1 to 0.08.

54 U=1/R.

55 McGee, C., Mosher, M. and Clarke, D. (2005) 'Insulation: Overview', in *Technical Manual: Design for Lifestyle and the Future* (third edition), Commonwealth of Australia, http://greenhouse.gov.au/yourhome/technical/fs16a.htm, accessed 21 July 2006.

56 McGee, C., Mosher, M. and Clarke, D. (2005) 'Insulation: Overview', in *Technical Manual: Design for Lifestyle and the Future* (third edition), Commonwealth of Australia, http://greenhouse.gov.au/yourhome/technical/fs16a.htm, accessed 21 July 2006.

57 See Energy Rating at http://search.energyrating.gov.au/air_srch.asp, accessed 21 July 2006.

58 Calculated as the fraction of the conventional solution's HVAC capital cost that corresponds to the decrease in design cooling load, q_{DES}.

59 Sustainable Energy Development Office (n.d) *Your Guide to Energy-Smart Air-Conditioners*, SEDO, Australia, www1.sedo.energy.wa.gov.au/uploads/Air-Conditioners_65.pdf, accessed 8 February 2006.

60 The large overhang modification is applied to the whole perimeter but only affects the windows on the northern, eastern and western façades.

61 Calculated using present values with an interest rate of 6 per cent, compounded annually.

62 Hawken, P., Lovins, A. B. and Lovins, L. H. (1999) *Natural Capitalism: Creating the Next Industrial Revolution*, Earthscan, London, pp87–88.

63 Hawken, P., Lovins, A. B. and Lovins, L. H. (1999) *Natural Capitalism: Creating the Next Industrial Revolution*, Earthscan, London, pp87–88.

10

Worked Example 5 – Domestic Water Systems

Significance of domestic water systems

Fresh water has been described as 'the most precious' natural resource in the world,[1] partly because it is vital for all living organisms and partly because less water is available for human use than most people realize. In fact, only about 0.3 per cent of all the free water on Earth is usable by humans,[2] as shown in Figure 10.1.

Many nations face water challenges in both urban and rural areas. The freshwater sources that supply many households are critically depleted due to a combination of non-sustainable extraction and drought. In response, stricter regulations, including regulations on household water consumption, are now being introduced to protect some of the remaining freshwater sources. In the long term, water consumers can expect the cost of freshwater to increase because of scarcity or because it is supplied from non-local sources.

In 2004–2005, total water consumption in Australia was 18,767GL – down 14 per cent from 2000–2001 due mainly to drought.[3] Figure 10.2 shows the division of water consumption within the Australian economy in 2004–2005.

The quantity of water consumed in Australian households is second only to the agricultural sector. In 2004–2005, water consumption in Australian households was 2,108GL (282 litres/person/day),[4] down 8 per cent from 2000–2001.[5] Figure 10.3 shows the quantity of water consumed from various sources. The high consumption of distributed water and low consumption of reused water suggests there may be opportunities to reuse distributed water in households.

The following worked example focuses cooling systems in houses. Specifically, it will explore the feasibility of household wastewater being cost-effectively treated and reused on-site in place of distributed water. It will also demonstrate how to optimize the whole domestic water system for multiple benefits, including cost savings.

Worked example overview

The on-site domestic water system is comprised of three categories of components.

1 Water-consuming appliances

Water-consuming appliances in a typical household are toilets, showers, baths, wash basins, sinks, dishwashers and washing machines. Once used, water becomes wastewater, which can be categorized as either greywater or blackwater. Typically, greywater is all domestic wastewater except for toilet wastewater, which is blackwater. The majority of domestic wastewater is greywater. Additional water consumption is from swimming pools, spas, and watering gardens and lawns. For the purpose of this worked example, swimming pools and spas are not considered and all irrigation requirements are assumed to be met by reusing the wastewater.

2 Wastewater treatment system

There are several types of wastewater treatment systems. They can be categorized by two characteristics:

1 The treatment stage (primary, secondary or tertiary); and
2 The treatment action (mechanical, chemical or biological).

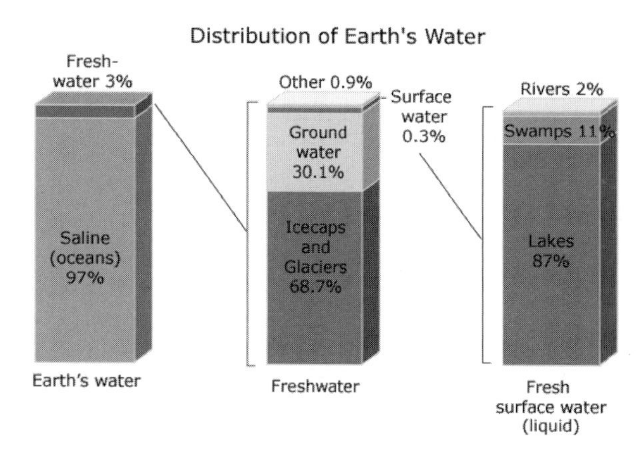

Source: US Geological Survey (2006)[6]

Figure 10.1 *Distribution of Earth's water*

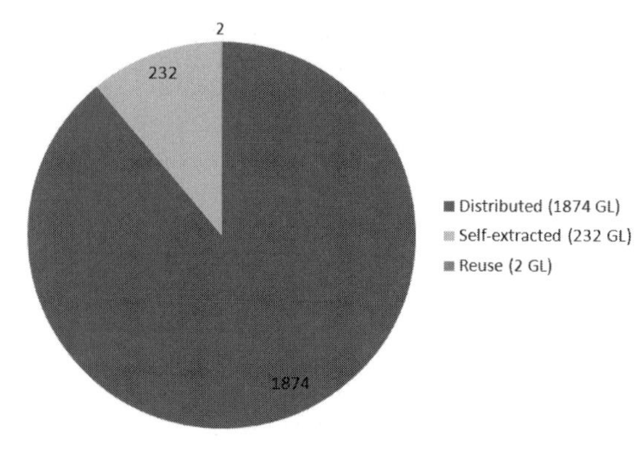

Source: Adapted from Trewin (2006), p103[9]

Figure 10.3 *Australian household water consumption in 2004–2005*[8]

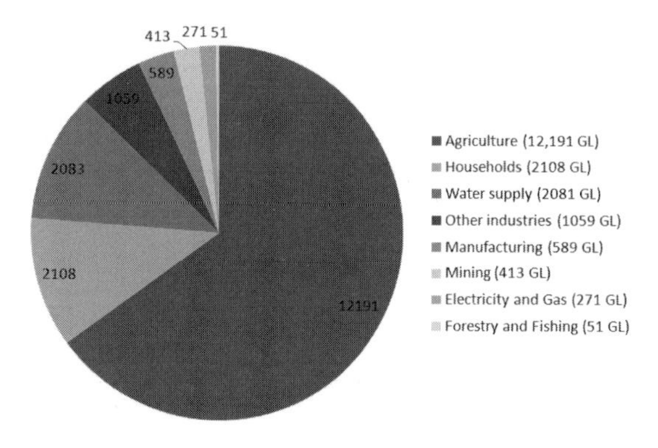

Source: Adapted from Trewin (2006), p8[7]

Note: Water Supply item includes sewerage and drainage services and losses.

Figure 10.2 *Australian water consumption in 2004–2005*

Table 10.1 indicates the treatment action generally used at each treatment stage. Primary and secondary wastewater treatment systems are considered in this worked example.

3 Discharge or reuse system

Treated wastewater can be discharged to on-site dispersal trenches or to nearby waterways. It can also be reused outdoors for irrigation or, in some cases, indoors as toilet water. The admissible dispersal and reuse systems vary with the wastewater treatment stage and local regulations. In this worked example wastewater is used for irrigation via a subsurface drip irrigation system.

Recall the elements of applying a Whole System Design (WSD) approach discussed in Chapters 4 and 5:

1 Ask the right questions;
2 Benchmark against the optimal system;
3 Design and optimize the whole system;
4 Account for all measurable impacts;
5 Design and optimize subsystems in the right sequence;
6 Design and optimize subsystems to achieve compounding resource savings;
7 Review the system for potential improvements;
8 Model the system;
9 Track technology innovation; and
10 Design to create future options.

The following worked example will demonstrate how the elements can be applied to domestic on-site water systems using two contrasting examples: a conventional domestic water system versus a WSD domestic water system. The application of an element will be indicated with a shaded box.

Table 10.1 *Wastewater treatment actions for each treatment stage*

Treatment stage	Treatment action	Description
Primary	Mechanical	Suspended solids are removed from the wastewater by settlement filtration.
Secondary	Biological	Organic materials are degraded by micro-organisms.
Tertiary	Mechanical Biological Chemical	Targeted inorganic nutrients and organic materials are removed by combinations of settlement, filtration, chemicals and micro-organisms, usually in a multistage process.

Design challenge

Design a domestic water system with on-site secondary wastewater treatment and wastewater reuse systems for a four bedroom house of five residents. Wastewater is from toilet, bathroom, kitchen and laundry appliances. The house is located in Adelaide, South Australia.

Design process

The following sections of this chapter present:

1 *Conventional design solution*: Conventional on-site system design with limited application of the elements of WSD;
2 *WSD solution*: Improved system using the elements of WSD; and
3 *Performance comparison*: Comparison of the economic and environmental costs and benefits.

Conventional design solution

The conventional solution incorporates:

- Standard water-consuming household appliances;
- An on-site wastewater treatment system: septic system and slow sand filter, as in Figure 10.4; and
- A water reuse system: subsurface drip irrigation system.

Standard water-consuming household appliances

Several appliances contribute to household water consumption and wastewater generation:

- Standard 9/4.5 litre dual flush toilets consume an average of 5.4L of water per flush[10] and cost about AU$300.
- Standard showerheads can consume 15–20L of water per minute[11] and cost about AU$50.[12]
- Standard taps discharge 15–18L of water per minute[13] and cost about AU$80.[14] Assume there are ten taps in the household.
- Average (Water Star Rating = 1.5) dishwashers consume about 19L of water per load[15] and cost about AU$700.[16]
- Average (Water Star Rating = 2) 8kg washing machines consume about 150L of water per load[17] and cost about AU$800.[18]

Cost

The standard water-consuming appliances' capital cost is:

$$\text{Capital cost} = \text{AU\$}300 + \text{AU\$}50 + (\text{AU\$}80)(10) + \text{AU\$}700 + \text{AU\$}800 = \text{AU\$}2650$$

The installation costs of standard appliances and water-efficient appliances (in the WSD solution) are comparable, so these costs are not considered.

Table 10.2 summarizes the water consumption of a household with standard appliances.

The standard appliances' water consumption in a household of five people is:

$$\text{Water consumption} = (150\text{L/person/day})(5\ \text{people})(91\ \text{days/quarter}) = 68{,}250\text{L/quarter}$$

This consumption is below the quarterly water consumption threshold set by the state water utility, and therefore not subject to excess-water-consumption rates. The regular rate in Australia is assumed to be AU$0.47/kL.

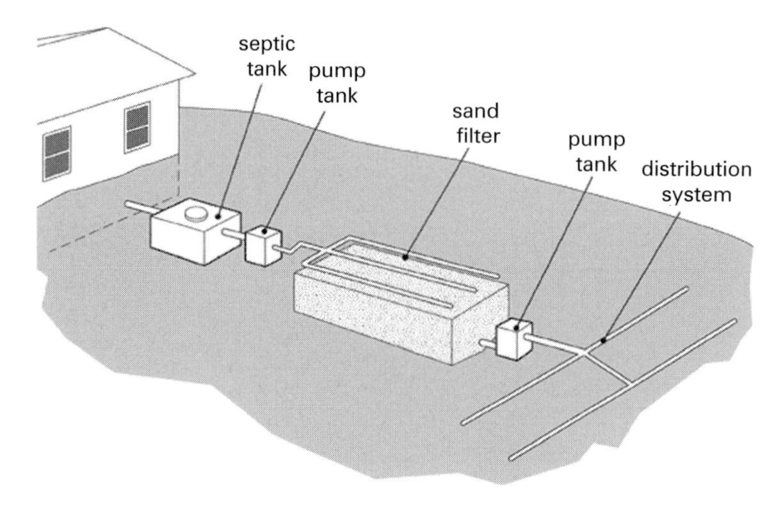

Source: Gustafson, Anderson and Heger Christopherson (2002)[19]

Figure 10.4 *Components of a conventional on-site wastewater treatment and reuse system*

Thus the total water cost is:

Water supply = (AU$0.47/kL)(68,250L/quarter)
(0.001L/kL)(4 quarters/year) = AU$129/year

Standard water-consuming appliances can last for at least 20 years if operated correctly.

On-site water treatment system: Septic system for primary treatment

Background

Septic systems are used for primary wastewater treatment and are available as both above-ground and below-ground systems. In a septic system, wastewater enters a tank, where it is treated and then it is discharged.

Figure 10.5 shows a typical septic tank. Wastewater enters the tank by active pumping or the action of gravity. The wastewater then settles and separates into three main zones. Materials such as grease, fats and oils that float to the top are called scum. Materials that settle to the bottom zone are called sludge. Between the scum and sludge is relatively clear water, although still containing bacteria and dissolved chemicals. The solids are anaerobically decomposed by bacteria in the tank.

The primary-treated water exits the tank by active pumping or the action of gravity and is then routed to a slow sand filter for secondary treatment before either being dispersed or used for irrigation.

Septic systems can process blackwater and greywater. However, their biological action can be impaired by household chemicals, gasoline, oil, pesticides, antifreeze and paint, which kill the bacteria that decompose the solids.[23] The system's flow can also be impaired by kitchen and bathroom items such as food wastes, toilet paper and sanitary items. The systems are not designed to process inorganic solids such as plastics and metals.

Table 10.2 *Daily water consumption for standard domestic appliances*

Waste source	Allowance (L/person/day)
Toilet	50
Bath and shower	50
Hand basin tap	10
Kitchen	10
Tap	7
Dishwasher	3
Laundry	30
Tap	5
Washing machine	25
Total	*150*[20]

Sources: NSW Health Department (2001), p12;[21] Ecological Homes (2002)[22]

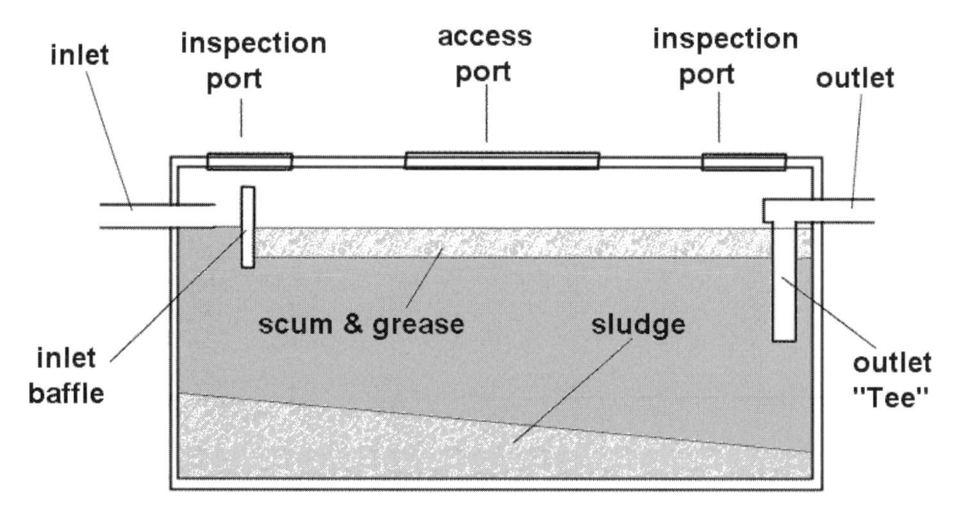

Source: SepticTankInfo[24]

Figure 10.5 *Cross-section of a single-compartment septic tank*

Septic systems function well at relatively steady loading, but their function can be impaired by heavy and shock loading. For example, the systems can handle one to two clothes washing loads, seven days per week, but cannot handle four or more in a single day. Excessive volumes of wastewater cause the grease, scum and sludge to mix with the water and escape. For this reason, septic systems are usually unable to process wastewater from high volume appliances such as hot tubs.

Cost

In Australia, there are standards and guidelines for septic systems at the federal, state and local levels of government. The system's capacity requirements vary from region to region. In South Australia, the South Australian Health Commission[25] requires that the septic tank capacity is at least:

Tank capacity = 3000L for up to 6 people
+ 1000L for each additional 2 people

Hence for this worked example:

Tank capacity = 3000L

Some Australian states require that the tank capacity is determined by the expected flow of wastewater. For example, the NSW Health Department[26] requires that the septic tank capacity is at least:

Tank capacity = sludge allowance + (daily water consumption)(number of people),[27]

where sludge allowance = 1550L, and 'daily water consumption' is as in Table 10.2 above. Hence:

Tank capacity = 1550L + (50 + 50 + 10 + 10 + 30L/person/day)(5 people) = 2300L

The system's capital cost depends on the components selected. The average cost for a septic system of suitable capacity, including a pump, is:

Capital cost = AU$4000[28]

The capital cost could vary by about AU$1000.

There are several extra costs involved in preparing septic systems for use, including delivery, excavation, installation, establishing electrical connections, quality checks, council approval and commissioning. These extra costs amount to:

Extra capital costs = AU$2500

These extra capital costs could vary by about AU$500. The system's running costs depend on the components selected and the loading volume and type:

- The power cost is approximately AU$25 per year, and could vary by about AU$5.[29]
- As with most other standards and guidelines, the South Australian Health Commission[30] suggests that septic systems that support five people need to be de-sludged approximately every four years. The actual need for de-sludging depends on the volume of solids in the tank, and so is subject to inspection. De-sludging costs approximately AU$300 per service and could vary by about AU$100.[31]
- Inspections are usually performed by trained personnel at a cost of approximately AU$70. Inspections are performed twice a year.
- Some components, such as the baffle, lid and pumps, may need to be replaced; this will incur costs. Wastewater can be highly corrosive and can damage internal components such as baffles.[32] Replacement parts costs are not considered here due to uncertainty.

Thus, the system's total running cost is:

Running cost = AU$25/year + AU$300/4years + (AU$70/service)(2 services/year) = AU$240/year

Septic systems can last for at least 20 years if built and operated correctly.[33] However, in practice, the systems usually last just a few years. More than 70 per cent of septic systems fail within eight years.[34] Thus the life of the septic system is estimated at ten years.

On-site water-treatment system: Slow sand filter system for secondary treatment

Background

Slow sand filter systems are used for secondary wastewater treatment. In a slow sand filter system, wastewater enters, is filtered and then dispersed. Figure 10.6 shows a typical slow sand filter. Septic tank effluent enters the tank by active pumping or the action of gravity.

The sand layer treats the effluent through physical and biological processes. The sand prevents suspended solids from passing through to the outlet. The sand also becomes coated by a thin biofilm,[36] which contains micro-organisms that decompose the organic matter and nutrients.[37] The biofilm usually develops in several days and is most prevalent in and above the top few centimetres of sand,[38] although it is present in about the top 40cm.[39] High-surface-area mediums other than sand are also used.

The gravel layer prevents sand moving to the outlet.[40] The gravel layer can be replaced by a geotextile layer, which is thinner and hence reduces the system's total height. The secondary-treated water exits the sump by active pumping or the action of gravity and is either dispersed or routed to an irrigation or reuse system.

Slow sand filter systems are best at processing suspended solids and bacteria in relatively clear wastewater. They cannot process heavy metals, chemicals

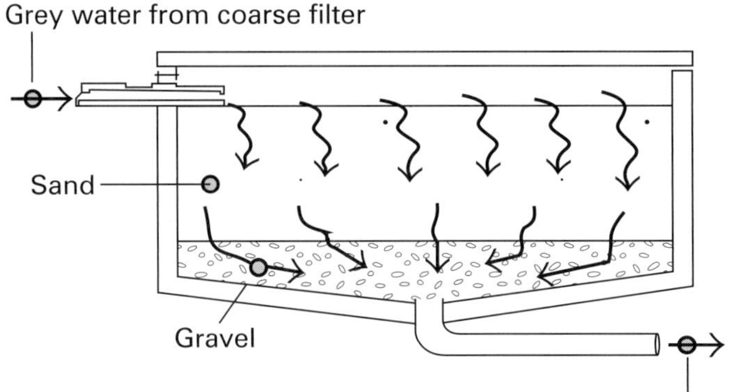

Fane and Reardon (2005)[35]

Figure 10.6 *Cross-section of a slow sand filter*

or other pollutants in excess.[41] Slow sand filter systems function well at relatively steady loading, but their slow action results in impaired function at heavy and shock loading. Consequently, slow sand filter systems may require an input control system, such as a timed pump.

Cost

In Australia, there are standards and guidelines for sand filter systems at the federal, state and local levels of government. The system's top surface area requirements vary from region to region. In this worked example the South Australian Health Commission[42] requires that the slow sand filter top surface area is at least:

Top surface area = the greater of:

1 Water consumption, = $1m^2$/50 L daily water consumption,

where 'daily water consumption' is as in Table 10.2 above, or

2 Organic load, = $1m^2$/25g BOD_5 daily organic load,

where 'daily organic load' is 50g BOD_5/person/day.[43] Hence,

Top surface area = the greater of:

1 Daily water consumption: = (5 people)(150L/person/day)/(50L) = $15m^2$ or

2 Daily organic loading: = (5 people)(50g BOD_5/person/day)/(25g BOD_5) = $10m^2$

Therefore,

Top surface area = $15m^2$

The average cost for a slow sand filter system of suitable capacity, including a pump, is:

Capital cost = AU$1000

There are several extra costs involved in preparing slow sand filter systems for use, including delivery, installation, quality checks, council approval and commissioning. These extra costs can amount to:

Extra capital costs = AU$500

The system's running costs depend on the components selected and the loading volume and type. The running costs are comprised of:

- Power cost for the pump,
- Maintenance and cleaning cost for the filter–the filter's effectiveness depends on a good wastewater flow rate. However, organic matter and silt can accumulate in the top layer of sand and restrict flow.[44] Consequently, the top layer of sand requires replacement about every six months and all of the sand requires replacement about every ten years;[45]
- Repair and replacement costs for the system's components.

Thus, the system's total running cost is:

Running cost = AU$400/year[46]

This running cost could vary by about AU$150/year.

Slow sand filter systems can last for at least 20 years if built and operated correctly.

Subsurface drip irrigation system

The capital cost of a subsurface drip irrigation system has been determined using Biolytix's online questionnaire.[47] The questionnaire results indicate that, for a system with standard appliances,[48] the 'Safe T Drip 400 Normal Flow' system[49] is suitable for the conventional solution. The system's capacity is determined by the wastewater volume.

Cost

The system's capital cost is:

Capital cost = AU$1336.20[50]

The system's installation cost is:

Installation cost = AU$1500

The system's running costs are comprised of:

- Pumping power costs, which are absorbed into the running costs of the slow sand filter system; and
- Inspection and maintenance costs, which are relatively high, particularly for labour.[51] Inspection for domestic systems is usually performed by the residents for no cost, thus inspection costs are not considered. Maintenance costs arise mainly from damage to the irrigation hosing by external events, such as piercing by shovels, and are rare. Hosing outlets are resistant to constriction by roots, so maintenance costs are not considered.

Subsurface drip irrigation systems can last for at least 20 years if operated correctly.

Whole system design solution

Element 7: Review the system for potential improvements

The WSD solution incorporates:

- Water-efficient household appliances;
- An on-site wastewater treatment system: Biolytix system; and
- A water reuse system: subsurface drip irrigation system.

Water-efficient appliances

Element 6: Design and optimize subsystems to achieve compounding resource savings

There are water-efficient models of every common household water-consuming appliance. The water-efficient appliances are usually more energy-efficient as well:

- Newer 6/3 litre toilets consume 3.6L per flush[52] and cost about AU$300.
- Water-efficient showerheads consume 6–7L per minute[53] and cost about AU$80.[54] Not only do efficient showerheads consume less water, they also reduce energy costs by 47 per cent due to the lower water heating demand.[55]
- Low-flow and aerating taps can discharge as little as 2L per minute[56] and cost about AU$100.[57] Assume there are 10 taps in the household.
- The most water-efficient (Water Star Rating = 4) dishwashers consume about 13L of water per load[58] and cost about AU$1000.[59]
- The most water-efficient (Water Star Rating = 5) 8kg washing machines consume about 60L of water per load[60] and cost about AU$1000.[61]

Cost

The water-efficient appliances' capital cost is:

Capital cost = AU$300 + AU$80 + (AU$100)(10) + AU$1000 + AU$1000 = AU$3380

The installation costs of standard appliances (in the conventional solution) and water-efficient appliances are comparable, so these costs are not considered.

Table 10.3 summarizes the water consumption of a household with water-efficient appliances.

The water-efficient appliances' water consumption in a home of five people is:

Water consumption = (5 people)(91 days/quarter) (64L/person/day) = 29,120L/quarter

This consumption is below the quarterly water consumption threshold set by the South Australian State water utility, so is not subject to excess-water-consumption rates. The regular rate in Australia is assumed to be AU$0.47/kL.

Thus, the annual water cost is:

Table 10.3 *Daily water consumption for water-efficient domestic appliances*

Waste source	Allowance (L/person/day)
Toilet	33
Bath and shower	19
Hand basin tap	1
Kitchen	3
Tap	1
Dishwasher	2
Laundry	11
Tap	1
Washing machine	10
Total	*64*

Water supply = (AU$0.47/kL)(29,120L/quarter)(0.001L/kL)(4 quarters/year) = AU$55/year

Water-efficient appliances can last for at least 20 years if operated correctly.

On-site water treatment system: Biolytix system for primary and secondary treatment

Background

> *Element 9: Track technology innovation*

Biolytix have filter systems for both primary and secondary wastewater treatment; these are available as both above-ground and below-ground systems. In a Biolytix system, wastewater enters the filter, where it is treated and then it is dispersed.

Figure 10.7 shows the filter for a Biolytix Deluxe system,[62] a secondary treatment system. Wastewater enters the tank by active pumping or gravity feed. The system also accepts solid materials such as food wastes and sanitary items.

The drainage layer houses a wet soil ecosystem consisting of organisms such as worms, beetles and micro-organisms. The organisms maintain the layer's porosity for good air circulation and drainage. They also decompose the wastewater and waste materials into humus, while the water and any remaining organic materials drain through to the humus layer. The humus layer also houses a soil ecosystem. The organisms reprocess the humus and organic materials into a fine, sponge-like matrix. The matrix, which is 90 per cent water by mass, has a high cation and anion exchange capacity, so it attracts and holds dissolved pollutants while the organisms decompose them. The water and any untreated solids drain through to the geotextile.

The geotextile filters out solids larger than 90 microns and is kept clean through biological action. The water drains through to the sump, where any remaining solids settle. The secondary-treated water exits the sump by active pumping or gravity feed and is then routed to an irrigation or reuse system. The Biolytix Rugged system, a primary treatment system, works by similar biological action. Like in the septic

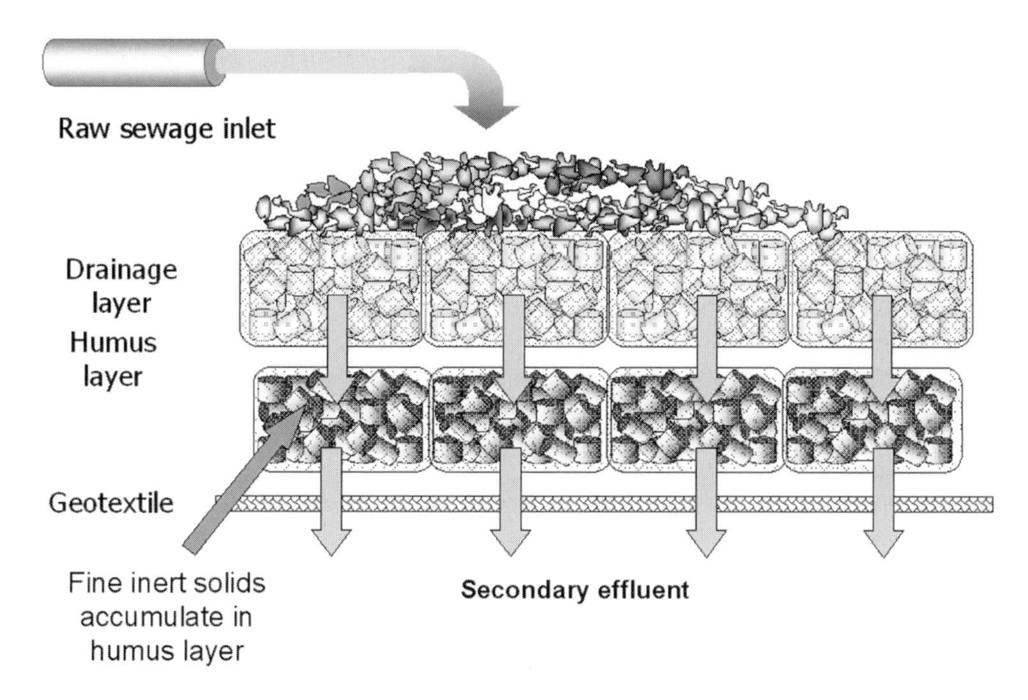

Source: Biolytix (2006f)[63]

Figure 10.7 *Cross-section of the Biolytix Deluxe system*

system, the primary-treated water can be dispersed or treated for irrigation.

Element 10: Create options for future generations

The Biolytix system can also be used to convert a septic system to a secondary wastewater treatment system.[64] Biolytix systems can process blackwater, greywater, and kitchen and bathroom items such as food wastes, sanitary items, paper, cardboard and household chemicals.[65] The systems are not designed to process inorganic solids such as plastics and metals. Biolytix systems function reliably at steady, heavy and shock loading.

Cost

Biolytix use an online questionnaire[66] to determine the suitable model and its cost. The questionnaire results indicate that a Biolytix Deluxe system (a secondary treatment system) incorporating the 'Pumped Audiovisual BF6_3000PAV' filter[67] is suitable for the WSD solution. The system includes an audio-visual alarm system that alerts Biolytix to disturbances and failures.

The system, which has the capacity to process the wastewater from ten people, or 1200L/day,[68] is the smallest capacity system currently available. However, for the design challenge, the system will be processing a sub-capacity volume of effluent:

Effluent volume = (64L/day/person)(5 people)
= 320L/day

The system's capital cost is:

Capital cost = AU$6329.10[69]

There are several extra costs involved in preparing Biolytix systems for use, including delivery, excavation, installation, establishing electrical connections, quality checks, council approval and commissioning. These extra costs amount to:

Extra capital costs = AU$2500

These extra capital costs could vary by about AU$500.

The system's running costs are:

- The power cost is approximately AU$15 per year, and could vary by about AU$5;[70] and
- Biolytix systems have an optional 20-year warranty that guarantees performance and component integrity.[71] The warranty covers the costs of all services, including removing excess humus once per year,[72] inspections, call-outs and component replacements.[73] The warranty cost for a Biolytix Pumped system is AU$352 per year.[74]

Thus, the system's total running cost is:

Running cost = AU$352/year + AU$15/year
= AU$367/year

Subsurface drip irrigation system

The capital cost of a subsurface drip irrigation system has been determined using Biolytix's online questionnaire.[75] The questionnaire results indicate that, for a system with water-efficient appliances,[76] the 'Safe T Drip 200 Normal Flow' system[77] is suitable. The system's capacity is determined by the wastewater volume.

Cost

The system's capital cost is:

Capital cost = AU$872.10[78]

The system's installation cost is:

Installation cost = AU$1000

The system's running costs are comprised of:

- Pumping power costs, which are absorbed into the running costs of the Biolytix system; and
- Inspection and maintenance costs, which are relatively high, particularly for labour.[79] Inspection for domestic systems is usually performed by the residents at no cost, so inspection costs are not considered. Maintenance costs arise mainly from damage to the irrigation hosing by external events, such as piercing by shovels, and are rare. Hosing

outlets are resistant to constriction by roots, so maintenance costs are not considered.

Subsurface drip irrigation systems can last for at least 20 years if operated correctly.

Summary: performance comparisons

A performance comparison reveals that while the capital costs of the conventional and WSD solutions are about equal, the long-term cost of the WSD solution is substantially lower.

Water-consuming appliances

In this worked example, the capital cost of water-efficient appliances is 28 per cent greater than that of standard appliances. However, water-efficient appliances consume less water and can thus reduce running costs if water is purchased, which is usually the case in Australia. In some rural or remote homes, water is not purchased, but collected during rainfall or pumped from underground. These alternative water sources can be limited, especially in the many parts of Australia where rainfall is infrequent. For these homes, low water consumption is not a cost saving but a necessity. Table 10.4 compares the capital costs, the water consumption and the running cost of the standard and water-efficient appliances. The results indicate that, in this worked example, the water-efficient appliances consume 57 per cent less water than standard appliances.

The water treatment and reuse system

The capital costs of the conventional and WSD water treatment and reuse systems are about equal. There is, however, a telling difference in the composition of the costs. The capital cost of the septic and slow sand filter systems together is AU$829 less than the Biolytix system, but this difference is roughly compensated for by the AU$964 lower cost for the lower capacity subsurface drip irrigation system in the WSD solution. The capacity of the subsurface drip irrigation system in the WSD solution is lower, because the water-efficient appliances reduce the wastewater volume. Furthermore, the life of a Biolytix system is, statistically, more than two times longer than a typical septic system. Consequently, there is an additional large investment of about AU$3630[80] for the conventional solution at ten years. The running costs of the WSD solution are lower, predominantly because there is only one pump not two, there are fewer moving parts that can fail, and removing humus is easier and cheaper than de-sludging or replacing sand. The running costs are actually likely to be even more in favour of the WSD solution, because replacement part costs for the septic system are not considered due to uncertainty. Table 10.5 compares the capital and running cost of the conventional and WSD water treatment and reuse systems.

Table 10.4 *Comparing the costs and water consumption of standard and water-efficient appliances*

Water-consuming appliances	Capital costs (not installed)	Water consumption	Running costs (water only)
Standard appliances	AU$2650	273kL/year	AU$129/year
Toilet	AU$300		
Shower head	AU$50		
Taps	AU$800		
Dishwasher	AU$700		
Washing machine	AU$800		
Water-efficient appliances	AU$3380	116kL/year	AU$55/year
Toilet	AU$300		
Shower head	AU$80		
Taps	AU$1000		
Dishwasher	AU$1000		
Washing machine	AU$1000		

Table 10.5 *Comparing the capital and running costs of the water treatment and reuse systems*

Water treatment and reuse system	Capital costs (installed)	Running costs
Conventional solution	AU$10,836	AU$640/year
Septic system	AU$6500	AU$240/year
Slow sand filter system	AU$1500	AU$400/year
Subsurface drip irrigation system	AU$2836	
WSD solution	AU$10,701	AU$367/year
Biolytix system	AU$8829	AU$367/year
Subsurface drip irrigation system	AU$1872	

Figures 10.8 and 10.9 summarize the component costs of the conventional and WSD solutions.

Table 10.6 and Figure 10.10 compare the total cost of the conventional and WSD solutions. The comparison is over a 20-year period with an interest rate of 6 per cent. The comparison assumes that the septic system is replaced after ten years and that water and electricity costs remain constant.[81] The 20-year cost of the WSD solution is 29 per cent less than that of the conventional solution. Figure 10.10 suggests that the WSD solution would still cost less if the septic system didn't need replacing at ten years. Table 10.6 also compares water consumption over 20 years. The WSD solution uses 57 per cent, or 3140kL, less than the conventional solution.[82]

Mutliple benefits

Element 4: Account for all measurable impacts

The WSD solution has several other benefits over the conventional solution:

- Water-efficient appliances that use hot water can also save on energy costs, since less hot water is heated.[83]
- Water-efficient appliances may have a longer useful life than standard water consuming appliances, due to less wear on components.
- The Biolytix system is substantially more compact than either the septic or slow sand filter systems[84]

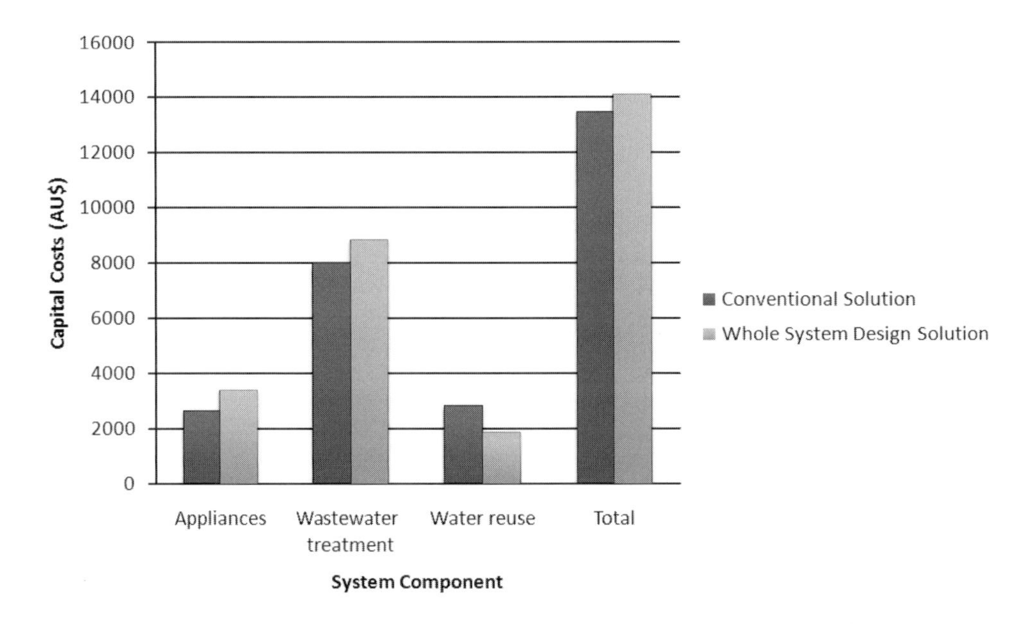

Figure 10.8 *Comparing the capital costs of components*

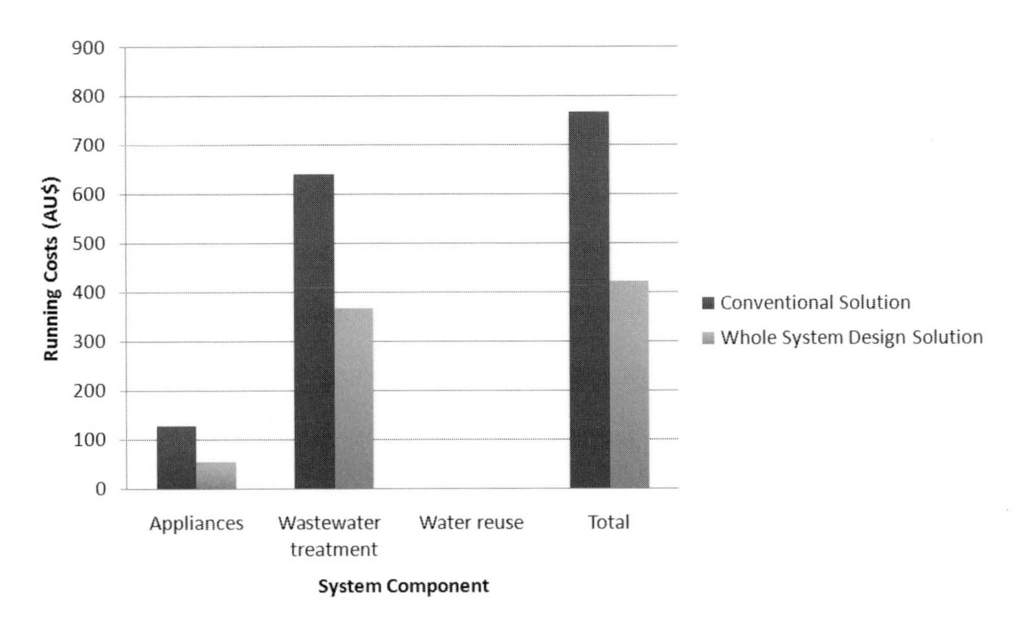

Figure 10.9 *Comparing the running costs of components*

Table 10.6 *Comparing the total cost of conventional and WSD systems over 20 years*

Solution	Capital costs	Running costs	20-year cost	20-year water consumption
Conventional	AU$13,486 + AU$3630 @ 10 years	AU$769/year	AU$25,741	5460kL
WSD	AU$14,081	AU$422/year	AU$18,311	2320kL

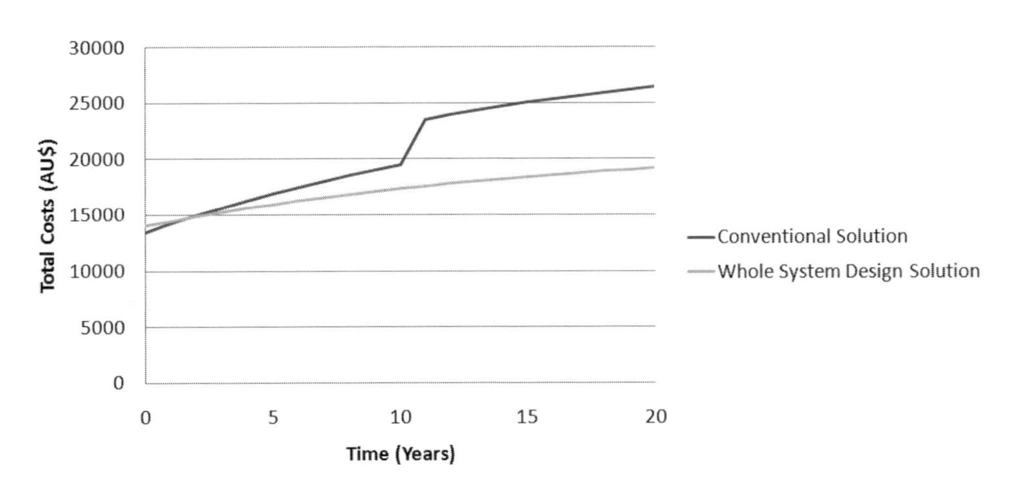

Figure 10.10 *Comparing the total cost of conventional and WSD systems over 20 years*

and, unlike the septic system, is not installed subsurface.

- There is no odour from the Biolytix system, even at high loading.[85]
- The Biolytix system can treat household chemicals and sanitary items, and handle heavy and shock loads, whereas the septic system can fail.[86]
- The Biolytix system does not produce greenhouse gases, whereas the septic system produces methane and hydrogen sulphide.[87]
- The Biolytix system requires one service per year,[88] whereas the septic and slow sand filter systems[89] require two inspections each.
- The subsurface drip irrigation system in the WSD solution requires less time for residents to inspect, because it is smaller.
- The subsurface drip irrigation system in the WSD solution is less likely to be damaged, because its surface coverage is smaller.

Notes

1 Carson, R. (1962) *Silent Spring*, Houghton Mifflin, Boston, MA.

2 US Geological Survey (2006) *Where is Earth's Water Located?*, US Department of the Interior, http://ga.water.usgs.gov/edu/earthwherewater.html, accessed 15 March 2007. This graph is derived from data in Gleick, P. H. (1996) 'Water resources', in S. H. Scheider (ed) *Encyclopedia of Climate and Weather*, Oxford University Press, New York, vol 2, pp817–823.

3 Trewin, D. (2006) '4610.0 Water account Australia 2004–2005', Australian Bureau of Statistics, Australia, p2, www.abs.gov.au/AUSSTATS/abs@.nsf/DetailsPage/4610.02004-05?OpenDocument, accessed 13 March 2007.

4 Calculated from 103kL per person.

5 Trewin, D. (2006) '4610.0 Water account Australia 2004–2005', Australian Bureau of Statistics, Australia, pp100-103, www.abs.gov.au/AUSSTATS/abs@.nsf/DetailsPage/4610.02004-05?OpenDocument, accessed 13 March 2007.

6 US Geological Survey (2006) *Where is Earth's Water Located?*, US Department of the Interior, http://ga.water.usgs.gov/edu/earthwherewater.html, accessed 15 March 2007. This graph is derived from data in Gleick, P. H. (1996) 'Water resources', in S. H. Scheider (ed) *Encyclopedia of Climate and Weather*, Oxford University Press, New York, vol 2, pp817–823.

7 Trewin, D. (2006) '4610.0 Water account Australia 2004–2005', Australian Bureau of Statistics, Australia, p8, www.abs.gov.au/AUSSTATS/abs@.nsf/DetailsPage/4610.02004-05?OpenDocument, accessed 13 March 2007.

8 Distributed sources are mains sources. Self-extracted sources include rainwater tanks and direct extraction from surface or groundwater.

9 Trewin, D. (2006) '4610.0 Water account Australia 2004–2005', Australian Bureau of Statistics, Australia, p103, www.abs.gov.au/AUSSTATS/abs@.nsf/DetailsPage/4610.02004-05?OpenDocument, accessed 13 March 2007.

10 Water Efficiency Labels and Standards Schemes (2007c) 'WELs products', Commonwealth of Australia, Australia, www.waterrating.gov.au/products/index.html, accessed 28 November 2006.

11 Water Efficiency Labels and Standards Schemes (2007c) 'WELs products', Commonwealth of Australia, Australia, www.waterrating.gov.au/products/index.html, accessed 28 November 2006.

12 ninemsn Shopping (2007) 'Taps at ninemsn Shopping', ninemsn Shopping, Australia, http://shopping.ninemsn.com.au/results/shp/?bCatId=2952, accessed 9 March 2007.

13 Water Efficiency Labels and Standards Schemes (2007c) 'WELs products', Commonwealth of Australia, Australia, www.waterrating.gov.au/products/index.html, accessed 28 November 2006.

14 ninemsn Shopping (2007) 'Taps at ninemsn Shopping', ninemsn Shopping, Australia, http://shopping.ninemsn.com.au/results/shp/?bCatId=2952, accessed 9 March 2007.

15 Water Efficiency Labels and Products Schemes (2007b) 'Dishwashers', Commonwealth of Australia, Australia, http://search.waterrating.com.au/dwashers_srch.asp, accessed 4 December 2006.

16 My Shopping (n.d.) 'Dishwashers', Comparison Shopping Australia, Australia, www.myshopping.com.au/PT--280_Dishwashers, accessed 9 March 2007; Shopping.com (2007) 'Dishwashers', Shopping.com Australia, Australia, http://au.shopping.com/xFA-dishwashers~FD-1894, accessed 9 March 2007.

17 Water Efficiency Labels and Products Schemes (2007a) 'Clothes washers', Commonwealth of Australia, Australia, http://search.waterrating.com.au/cwashers_srch.asp, accessed 11 January 2007.

18 My Shopping (n.d.) 'Washing machines', Comparison Shopping Australia, Australia, www.myshopping.com.au/PT--281_Washing_Machines, accessed 9 March 2007; Shopping.com (2007) 'Washing machines', Shopping.com Australia, Australia, http://au.shopping.com/xFA-washing_machines~FD-1897, accessed 9 March 2007.

19 Gustafson, D. M., Anderson, J. L. and Heger Christopherson, S. (2002) 'Innovative onsite sewerage treatment systems: Single-pass sand filters', Regents of the University of Minnesota Extension, US,

www.extension.umn.edu/distribution/naturalresources/DD7672.html, accessed 8 February 2007.

20 This figure is substantially lower than the average Australian household water consumption of 282L/person/day.

21 NSW Health Department (2001) 'Septic tank and collection well accreditation guideline', New South Wales State Government, Australia, www.health.nsw.gov.au/public-health/ehb/general/wastewater/septic_guideline.pdf, accessed 28 November 2006.

22 Ecological Homes (2002) 'Wastewater systems', *Ecological Homes*, Australia, www.ecologicalhomes.com.au/wastewater_systems.htm, accessed 28 November 2006.

23 US Office of Water (2005) *A Homeowner's Guide to Septic Systems*, US Environmental Protection Agency, p8, www.epa.gov/owm/septic/pubs/homeowner_guide_long.pdf, accessed 30 November 2006.

24 See SepticTankInfo - 'Septic tanks and septic systems' at http://septictankinfo.com/septic_tank_basics.shtml, accessed 29 November 2006.

25 Public and Environmental Health Service (1995) 'Waste control systems: Standard for the construction, installation and operation of septic tank systems in South Australia', South Australian Health Commission, Australia, p12, www.dh.sa.gov.au/pehs/publications/Septic-tank-book.pdf, accessed 28 November 2006.

26 NSW Health Department (2001) 'Septic tank and collection well accreditation guideline', New South Wales State Government, Australia, p12, www.health.nsw.gov.au/public-health/ehb/general/wastewater/septic_guideline.pdf, accessed 28 November 2006.

27 Number of people must be between five and ten.

28 Mark Quan, Icon Septec, personal communication on 13 February 2007; Steve Little, Steve Little Plumbing, personal communication on 13 February 2007; Palmer et al (2001) cited in Coombes, P. (2002) 'Water Smart Practice Note 9: Wastewater reuse', Lower Hunter and Central Coast Regional Environment Management Strategy, Australia, p4, www.portstephens.nsw.gov.au/files/51064/File/9_Wastewater.pdf, accessed 28 November 2006; Septreat, personal communication on 15 February 2007.

29 Waterpac Plumbing (2002) 'Getting started...', Waterpac Plumbing, Australia, www.waterpacaustralia.com/getting_started_.htm, accessed 28 November 2006; Hankinson, M. (2005) 'Bodalla sewerage: Community Newsletter No 2', *Eurobodalla Shire Council Newsletter*, Australia, p3, www.esc.nsw.gov.au/IWCMP/newsletters/BodallaNewsletter_No2.pdf, accessed 17 November 2006.

30 Public and Environmental Health Service (1995) 'Waste control systems: Standard for the construction, installation and operation of septic tank systems in South Australia', South Australian Health Commission, Australia, p13, www.dh.sa.gov.au/pehs/publications/Septic-tank-book.pdf, accessed 28 November 2006.

31 Hankinson, M. (2005) 'Bodalla sewerage: Community Newsletter No 2', *Eurobodalla Shire Council Newsletter*, Australia, p3, www.esc.nsw.gov.au/IWCMP/newsletters/BodallaNewsletter_No2.pdf, accessed 12 November 2006; Biolytix, personal communication on 13 February 2007.

32 Ward, R. C. and Englehardt, J. D. (1993) *Management of Decentralized, On-Site Systems for Treatment of Domestic Wastes*, Purdue Research Foundation, US, www.purdue.edu/dp/envirosoft/decent/src/title.htm, accessed 28 November 2006.

33 The Laundry Alternative Inc. (2005) 'Septic system price', The Laundry Alternative Inc., US, www.laundry-alternative.com/septic_system_price.htm, accessed 28 November 2006.

34 Biolytix website – 'Competitor comparison', Biolytix, Australia, www.biolytix.com/detail.php?ID=76, accessed 28 November 2006.

35 Fane, S. and Reardon, C. (2005) '2.3 Wastewater reuse', in *Your Home Technical Manual* (third edition), Commonwealth of Australia, www.yourhome.gov.au/technical/fs23.htm, accessed 25 March 2008.

36 Cooperative Research Centre for Water Quality and Treatment (2006) *A Consumer's Guide to Drinking Water*, CRC for Water Quality, Australia, p26, www.waterquality.crc.org.au/consumers/consumer.pdf, accessed 11 February 2007.

37 Gustafson, D. M., Anderson, J. L. and Heger Christopherson, S. (2002) 'Innovative onsite sewerage treatment systems: Single-pass sand filters', Regents of the University of Minnesota Extension, US, www.extension.umn.edu/distribution/naturalresources/DD7672.html, accessed 8 February 2007.

38 Oasis Design for AWWA Research Foundation (1999) *Slow Sand Filtration*, Oasis Design for AWWA Research Foundation, US, www.oasisdesign.net/water/treatment/slowsandfilter.htm, accessed 7 February 2007.

39 Fox, R. (1995) 'Slow sand filtration', *Practical Hydroponics and Greenhouses*, no 24, Casper Publications, Australia, www.hydroponics.com.au/back_issues/issue24.html, accessed 7 February 2007.

40 Oasis Design for AWWA Research Foundation (1999) *Slow Sand Filtration*, Oasis Design for AWWA Research Foundation, US, www.oasisdesign.net/water/treatment/slowsandfilter.htm, accessed 7 February 2007.

41 Oasis Design for AWWA Research Foundation (1999) *Slow Sand Filtration*, Oasis Design for AWWA Research Foundation, US, www.oasisdesign.net/water/treatment/slowsandfilter.htm, accessed 7 February 2007.

42 Public and Environmental Health Service (1998) 'Waste control systems: Standard for the construction, installation and operation of septic tank systems in

South Australia: Supplement A – Aerobic sand filters', South Australian Health Commission, Australia, p1, www.dh.sa.gov.au/pehs/publications/Supplement-A.pdf, accessed 28 November 2006.

43 BOD_5 = the five-day Biological Oxygen Demand.

44 Oasis Design for AWWA Research Foundation (1999) *Slow Sand Filtration*, Oasis Design for AWWA Research Foundation, US, www.oasisdesign.net/water/treatment/slowsandfilter.htm, accessed 7 February 2007.

45 Gustafson, D. M., Anderson, J. L. and Heger Christopherson, S. (2002) 'Innovative onsite sewerage treatment systems: Single-pass sand filters', Regents of the University of Minnesota Extension, US, www.extension.umn.edu/distribution/naturalresources/DD7672.html, accessed 8 February 2007.

46 Gustafson, D. M., Anderson, J. L. and Heger Christopherson, S. (2002) 'Innovative onsite sewerage treatment systems: Single-pass sand filters', Regents of the University of Minnesota Extension, US, www.extension.umn.edu/distribution/naturalresources/DD7672.html, accessed 8 February 2007.

47 Biolytix website – 'Welcome to the Biolytix product selection wizard', Biolytix, Australia, www.biolytix.com/php/productSelection, accessed 28 November 2006.

48 Approximated as 'Standard plumbing fixtures' in the Biolytix online questionnaire.

49 Biolytix website – 'Drip irrigation kits', Biolytix, Australia, www.biolytix.com/detail.php?ID=90, accessed 12 February 2007.

50 Biolytix website – 'Drip irrigation kits', Biolytix, Australia, www.biolytix.com/detail.php?ID=90, accessed 12 February 2007.

51 Qassim, A. (2003) *Subsurface Irrigation: A Situation Analysis*, Department of Primary Industries, Victorian State Government, Australia, www.dpi.vic.gov.au/dpi/nrenfa.nsf/93a98744f6ec41bd4a256c8e00013aa9/3d3915fb8fe0af31ca256eb4001e5bf1/$FILE/Subsurface%20Irrigation.pdf, accessed 12 February 2007.

52 Water Efficiency Labels and Standards Schemes (2007c) 'WELs products', Commonwealth of Australia, Australia, www.waterrating.gov.au/products/index.html, accessed 28 November 2006.

53 Water Efficiency Labels and Standards Schemes (2007c) 'WELs products', Commonwealth of Australia, Australia, www.waterrating.gov.au/products/index.html, accessed 28 November 2006.

54 ninemsn Shopping (2007) 'Taps at ninemsn Shopping', ninemsn Shopping, Australia, http://shopping.ninemsn.com.au/results/shp/?bCatId=2952, accessed 9 March 2007.

55 Water Efficiency Labels and Standards Schemes (2007c) 'WELs products', Commonwealth of Australia, Australia, www.waterrating.gov.au/products/index.html, accessed 28 November 2006.

56 Water Efficiency Labels and Standards Schemes (2007c) 'WELs products', Commonwealth of Australia, Australia, www.waterrating.gov.au/products/index.html, accessed 28 November 2006.

57 ninemsn Shopping (2007) 'Taps at ninemsn Shopping', ninemsn Shopping, Australia, http://shopping.ninemsn.com.au/results/shp/?bCatId=2952, accessed 9 March 2007.

58 Water Efficiency Labels and Products Schemes (2007b) 'Dishwashers', Commonwealth of Australia, Australia, http://search.waterrating.com.au/dwashers_srch.asp, accessed 4 December 2006.

59 My Shopping (n.d.a) 'Dishwashers', Comparison Shopping Australia, Australia, www.myshopping.com.au/PT--280_Dishwashers, accessed 9 March 2007; Shopping.com (2007) 'Dishwashers', Shopping.com Australia, Australia, http://au.shopping.com/xFA-dishwashers~FD-1894, accessed 9 March 2007.

60 Water Efficiency Labels and Products Schemes (2007a) 'Clothes washers', Commonwealth of Australia, Australia, http://search.waterrating.com.au/cwashers_srch.asp, accessed 11 January 2007.

61 My Shopping (n.d.b) 'Washing machines', Comparison Shopping Australia, Australia, www.myshopping.com.au/PT--281_Washing_Machines, accessed 9 March 2007; Shopping.com (2007) 'Washing machines', Shopping.com Australia, Australia, http://au.shopping.com/xFA-washing_machines~FD-1897, accessed 9 March 2007.

62 Biolytix website – 'How Biolytix works', Biolytix, Australia, www.biolytix.com/detail.php?ID=69, accessed 28 November 2006.

63 Biolytix website – 'How Biolytix works', Biolytix, Australia, www.biolytix.com/detail.php?ID=69, accessed 28 November 2006.

64 Biolytix website – 'Competitor comparison', Biolytix, Australia, www.biolytix.com/detail.php?ID=76, accessed 28 November 2006.

65 Biolytix website – 'Competitor comparison', Biolytix, Australia, www.biolytix.com/detail.php?ID=76, accessed 28 November 2006.

66 Biolytix website – 'Welcome to the Biolytix product selection wizard', Biolytix, Australia, www.biolytix.com/php/productSelection, accessed 28 November 2006.

67 Biolytix website – 'Biolytix Filter Deluxe products (secondary treatment)', Biolytix, Australia, www.biolytix.com/detail.php?ID=27, accessed 10 February 2007.

68 Biolytix website – 'Info Kit', Biolytix, Australia, p10, www.biolytix.com/docs/Biolytixinfokit.pdf, accessed 4 December 2006.

69 Biolytix website – 'Biolytix Filter Deluxe products (secondary treatment)', Biolytix, Australia, www.biolytix.com/detail.php?ID=27, accessed 10 February 2007.

70 Biolytix website – *Household product range*, Biolytix, Australia, www.biolytix.com/detail.php?ID=11, accessed

28 November 2006; Biolytix, personal communication on 13 February 2007.

71 Biolytix website – 'Biolytix delivers the best service', Biolytix, Australia, www.biolytix.com/detail.php?ID=57, accessed 28 November 2006.

72 Biolytix website – 'Competitor comparison', Biolytix, Australia, www.biolytix.com/detail.php?ID=76, accessed 28 November 2006.

73 Biolytix website – 'Biolytix delivers the best service', Biolytix, Australia, www.biolytix.com/detail.php?ID=57, accessed 28 November 2006.

74 Biolytix, personal communication on 13 February 2007.

75 Biolytix website – 'Welcome to the Biolytix product selection wizard', Biolytix, Australia, www.biolytix .com/ php/productSelection, accessed 28 November 2006.

76 Approximated as 'Full water conservation fixtures' in the Biolytix online questionnaire.

77 Biolytix website – 'Drip irrigation kits', Biolytix, Australia, www.biolytix.com/detail.php?ID=90, accessed 12 February 2007.

78 Biolytix website – 'Drip irrigation kits', Biolytix, Australia, www.biolytix.com/detail.php?ID=90, accessed 12 February 2007.

79 Qassim, A. (2003) *Subsurface Irrigation: A Situation Analysis*, Department of Primary Industries, Victorian State Government, Australia, www.dpi.vic.gov.au/dpi/nrenfa.nsf/93a98744f6ec41bd4a256c8e00013aa9/3d3915fb8fe0af31ca256eb4001e5bf1/$FILE/Subsurface%20Irrigation.pdf, accessed 12 February 2007.

80 This value represents the net present value of an AU$6500 investment in 10 years at an interest rate of 6 per cent.

81 Assuming that water and energy costs remain constant over 20 years is unlikely. Water and energy costs are likely to increase and hence the total cost becomes more favourable for the WSD solution.

82 The actual water consumption over 20 years is likely to be higher for both the conventional and WSD solutions, because the performance of the appliances is likely to decrease.

83 Water Efficiency Labels and Standards Schemes (2007c) 'WELs products', Commonwealth of Australia, Australia, www.waterrating.gov.au/products/index.html, accessed 28 November 2006.

84 Biolytix website – 'Competitor comparison', Biolytix, Australia, www.biolytix.com/detail.php?ID=76, accessed 28 November 2006.

85 Biolytix website – 'Competitor comparison', Biolytix, Australia, www.biolytix.com/detail.php?ID=76, accessed 28 November 2006.

86 Biolytix website – 'Competitor comparison', Biolytix, Australia, www.biolytix.com/detail.php?ID=76, accessed 28 November 2006.

87 Biolytix website – 'Competitor comparison', Biolytix, Australia, www.biolytix.com/detail.php?ID=76, accessed 28 November 2006.

88 Biolytix website – 'Competitor comparison', Biolytix, Australia, www.biolytix.com/detail.php?ID=76, accessed 28 November 2006.

89 Gustafson, D. M., Anderson, J. L. and Heger Christopherson, S. (2002) 'Innovative onsite sewerage treatment systems: Single-pass sand filters', Regents of the University of Minnesota Extension, US, www.extension.umn.edu/distribution/naturalresources/DD7672.html, accessed 8 February 2007.

Index